Praise for David Eagleman's

Livewired

"Delivers an intellectually exhilarating look at neuroplasticity. . . . Eagleman's skill as teacher, bold vision, and command of current research will make this superb work a curious reader's delight."
— *Publishers Weekly* (starred review)

"Eagleman brings the subject to life in a way I haven't seen other writers achieve before." —Clare Wilson, *New Scientist*

"[Eagleman's] knowledge and enthusiasm are intoxicating. His book demonstrates the principle about which he is writing; my mind has been changed by his words." —Russell Brand

"The pages of *Livewired* are chock-full of mind-bending ideas and dazzling insights. Eagleman's infectious enthusiasm, his use of fascinating anecdotes, and his clear, effortless prose render the secrets of the brain's adaptability into a truly compelling page-turner. *Livewired* is a fun and whirlwind exploration of the most complex thing in the universe." —Khaled Hosseini,
author of *The Kite Runner*

"David's a brilliant writer and thinker, and he knows more about how we tick and why we tick than anyone I know." —Neil Gaiman

David Eagleman

Livewired

Dr. David Eagleman is a neuroscientist and internationally bestselling author. He teaches brain plasticity at Stanford University, is the creator and host of the Emmy-nominated television series *The Brain*, and is the CEO of Neosensory, a company that builds the next generation of neuroscience hardware. The author of seven previous books, Eagleman lives in Silicon Valley in California.

www.eagleman.com

Livewired

THE INSIDE STORY
OF THE EVER-CHANGING BRAIN

David Eagleman

VINTAGE BOOKS

A DIVISION OF PENGUIN RANDOM HOUSE LLC

NEW YORK

FIRST VINTAGE BOOKS EDITION, MAY 2021

Copyright © 2020 by David Eagleman

All rights reserved. Published in the United States by Vintage Books, a division of Penguin Random House LLC, New York. Originally published in hardcover in the United States by Pantheon Books, a division of Penguin Random House LLC, New York, in 2020.

Vintage and colophon are registered trademarks of Penguin Random House LLC.

The Library of Congress has cataloged the Pantheon edition as follows:
Name: Eagleman, David, author.
Title: Livewired : the inside story of the ever-changing brain / David Eagleman.
Description: First edition. | New York : Pantheon Books, 2020. | Includes
    bibliographical references and index.
Identifiers: LCCN 2020000081
Subjects: LCSH: Brain. | Neuroplasticity. | Learning—Physiological aspects.
Classification: LCC QP376 .E254 2020 | DDC 612.8/2—dc23
LC record available at https://lccn.loc.gov/2020000081

Vintage Books Trade Paperback ISBN: 978-0-307-94969-1
eBook ISBN: 978-0-307-90750-9

Author photograph © Mark Clark
Book design by Maggie Hinders

www.vintagebooks.com

Printed in the United States of America
10   9   8   7   6   5

CONTENTS

Livewired

Every man is born as many men and dies as a single one.

—MARTIN HEIDEGGER

1

THE ELECTRIC LIVING FABRIC

Imagine this: instead of sending a four-hundred-pound rover vehicle to Mars, we merely shoot over to the planet a single sphere, one that can fit on the end of a pin. Using energy from sources around it, the sphere divides itself into a diversified army of similar spheres. The spheres hang on to each other and sprout features: wheels, lenses, temperature sensors, and a full internal guidance system. You'd be gobsmacked to watch such a system discharge itself.

But you only need to go to any nursery to see this unpacking in action. You'll see wailing babies who began as a single, microscopic, fertilized egg and are now in the process of emancipating themselves into enormous humans, replete with photon detectors, multi-jointed appendages, pressure sensors, blood pumps, and machinery for metabolizing power from all around them.

But this isn't even the best part about humans; there's something more astonishing. Our machinery isn't fully preprogrammed, but instead shapes itself by interacting with the world. As we grow, we constantly rewrite our brain's circuitry to tackle challenges, leverage opportunities, and understand the social structures around us.

Our species has successfully taken over every corner of the globe because we represent the highest expression of a trick that Mother Nature discovered: don't entirely pre-script the brain; instead, just set it up with the basic building blocks and get it into the world. The bawling baby eventually stops crying, looks around, and absorbs the world around it. It molds itself to the surroundings. It soaks up everything from local language to broader culture to global politics. It carries forward the beliefs and biases of those who raise it. Every fond memory it possesses, every lesson it learns, every drop of information it drinks— all these fashion its circuits to develop something that was never preplanned, but instead reflects the world around it.

This book will show how our brains incessantly reconfigure their own wiring, and what that means for our lives and our futures. Along the way, we'll find our story illuminated by many questions: Why did people in the 1980s (and only in the 1980s) see book pages as slightly red? Why is the world's best archer armless? Why do we dream each night, and what does that have to do with the rotation of the planet? What does drug withdrawal have in common with a broken heart? Why is the enemy of memory not time but other memories? How can a blind person learn to see with her tongue or a deaf person learn to hear with his skin? Might we someday be able to read the rough details of someone's life from the microscopic structure etched in their forest of brain cells?

THE CHILD WITH HALF A BRAIN

While Valerie S. was getting ready for work, her three-year-old son, Matthew, collapsed on the floor.[1] He was unarousable. His lips turned blue.

Valerie called her husband in a panic. "Why are you calling me?" he bellowed. "Call the doctor!"

A trip to the emergency room was followed by a long aftermath of appointments. The pediatrician recommended Matthew have his heart checked. The cardiologist outfitted him with a heart monitor, which Matthew kept unplugging. All the visits surfaced nothing in particular. The scare was a one-off event.

Or so they thought. A month later, while he was eating, Matthew's face took on a strange expression. His eyes became intense, his right arm stiffened and straightened up above his head, and he remained unresponsive for about a minute. Again Valerie rushed him to the doctors; again there was no clear diagnosis.

Then it happened again the next day.

A neurologist hooked up Matthew with a cap of electrodes to measure his brain activity, and that's when he found the telltale signs of epilepsy. Matthew was put on seizure medications.

The medications helped, but not for long. Soon Matthew was having a series of intractable seizures, separated from one another first by an hour, then by forty-five minutes, then by thirty minutes—like the shortening durations between a woman's contractions during labor. After a time he was suffering a seizure every two minutes. Valerie and her husband, Jim, hurried Matthew to the hospital each time such a series began, and he'd be housed there for days to weeks. After several stints of this routine, they would wait until his "contractions" had reached the twenty-minute mark and then call ahead to the hospital, climb in the car, and get Matthew something to eat at McDonald's on the way there.

Matthew, meanwhile, labored to enjoy life between seizures.

The family checked into the hospital ten times each year. This routine continued for three years. Valerie and Jim began to mourn the loss of their healthy child—not because he was going to die, but because he was no longer going to live a normal life. They went through anger and denial. Their normal changed. Finally, during a three-week hospital stay, the neurologists had to allow that this problem was bigger than they knew how to handle at the local hospital.

So the family took an air ambulance flight from their home in Albuquerque, New Mexico, to Johns Hopkins hospital in Baltimore. It was here, in the pediatric intensive care unit, that they came to understand that Matthew had Rasmussen's encephalitis, a rare, chronic inflammatory disease. The problem with the disease is that it affects not just a small bit of the brain but an entire half. Valerie and Jim explored their options and were alarmed to learn there was only one known treatment for Matthew's condition: a hemispherectomy, or the surgical removal of an entire half of the brain. "I can't tell you anything the doctors said after that," Valerie told me. "One just shuts down, like everyone's talking a foreign language."

Valerie and Jim tried other approaches, but they proved fruitless. When Valerie called Johns Hopkins hospital to schedule the hemispherectomy some months later, the doctor asked her, "Are you sure?"

"Yes," she said.

"Can you look in the mirror every day and know you've chosen what you've needed to do?"

Valerie and Jim couldn't sleep beneath the crushing anxiety. Could Matthew survive the surgery? Was it even possible to live with half of the brain missing? And even if so, would the removal of one hemisphere be so debilitating as to offer Matthew a life on terms not worth taking?

But there were no more options. A normal life couldn't be lived in the shadow of multiple seizures each day. They found themselves weighing Matthew's assured disadvantages against an uncertain surgical outcome.

Matthew's parents flew him to the hospital in Baltimore. Under a small child-sized mask, Matthew drifted away into the anesthesia. A blade carefully opened a slit in his shaved scalp. A bone drill cut a circular burr hole in his skull.

Working patiently over the course of several hours, the surgeon removed half of the delicate pink material that underpinned Matthew's intellect, emotion, language, sense of humor, fears, and loves. The extracted brain tissue, useless outside its biological milieu, was

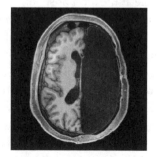

Half of Matthew's brain was surgically removed.

banked in small containers. The empty half of Matthew's skull slowly filled up with cerebrospinal fluid, appearing in neuroimaging as a black void.[2]

In the recovery room, his parents drank hospital coffee and waited for Matthew to open his eyes. What would their son be like now? Who would he be with only half a brain?

———

Of all the objects our species has discovered on the planet, nothing rivals the complexity of our own brains. The human brain consists of eighty-six billion cells called neurons: cells that shuttle information rapidly in the form of traveling voltage spikes.[3] Neurons are densely connected to one another in intricate, forest-like networks, and the total number of connections between the neurons in your head is in the hundreds of trillions (around 0.2 quadrillion). To calibrate yourself, think of it this way: there are twenty times more connections in a cubic millimeter of cortical tissue than there are human beings on the entire planet.

But it's not the number of parts that make a brain interesting; it's the way those parts interact.

In textbooks, media advertisements, and popular culture, the brain is typically portrayed as an organ with different regions dedicated to specific tasks. This area here exists for vision, that swath there is necessary for knowing how to use tools, this region becomes active when resisting candy, and that spot lights up when mulling over a moral conundrum. All the areas can be neatly labeled and categorized.

But that textbook model is inadequate, and it misses the most interesting part of the story. The brain is a dynamic system, constantly altering its own circuitry to match the demands of the environment and the capabilities of the body. If you had a magical video camera with which to zoom in to the living, microscopic cosmos inside the skull, you would witness the neurons' tentacle-like extensions grasping around, feeling, bumping against one another, searching for the right connections to form or forgo, like citizens of a country establishing friendships, marriages, neighborhoods, political parties, vendettas, and social networks. Think of the brain as a living community of trillions of intertwining organisms. Much stranger than the textbook picture, the brain is a cryptic kind of computational material, a living three-dimensional textile that shifts, reacts, and adjusts itself to maximize its efficiency. The elaborate pattern of connections in the brain— the circuitry—is full of life: connections between neurons ceaselessly blossom, die, and reconfigure. You are a different person than you were at this time last year, because the gargantuan tapestry of your brain has woven itself into something new.

When you learn something—the location of a restaurant you like, a piece of gossip about your boss, that addictive new song on the radio—your brain physically changes. The same thing happens when you experience a financial success, a social fiasco, or an emotional awakening. When you shoot a basketball, disagree with a colleague, fly into a new city, gaze at a nostalgic photo, or hear the mellifluous tones of a beloved voice, the immense, intertwining jungles of your brain work themselves into something slightly different from what they were a moment before. These changes sum up to our memories: the outcome of our living and loving. Accumulating over minutes and months and decades, the innumerable brain changes tally up to what we call you.

Or at least the you right now. Yesterday you were marginally different. And tomorrow you'll be someone else again.

LIFE'S OTHER SECRET

In 1953, Francis Crick burst into the Eagle and Child pub. He announced to the startled swillers that he and James Watson had just discovered the secret of life: they had deciphered the double-helical structure of DNA. It was one of the great pub-crashing moments of science.

But it turns out that Crick and Watson had discovered only *half* the secret. The other half you won't find written in a sequence of DNA base pairs, and you won't find it written in a textbook. Not now, not ever.

Because the other half is all around you. It is every bit of experience you have with the world: the textures and tastes, the caresses and car accidents, the languages and love stories.[4]

To appreciate this, imagine you were born thirty thousand years ago. You have exactly your same DNA, but you slide out of the womb and open your eyes onto a different time period. What would you be like? Would you relish dancing in pelts around the fire while marveling at stars? Would you bellow from a treetop to warn of approaching saber-toothed tigers? Would you be anxious about sleeping outdoors when rain clouds bloomed overhead?

Whatever you think you'd be like, you're wrong. It's a trick question.

Because you wouldn't be you. Not even vaguely. This caveman with identical DNA might *look* a bit like you, as a result of having the same genomic recipe book. But the caveman wouldn't think like you. Nor would the caveman strategize, imagine, love, or simulate the past and future quite as you do.

Why? Because the caveman's experiences are different from yours. Although DNA is a part of the story of your life, it is only a small part. The rest of the story involves the rich details of your experiences and your environment, all of which sculpt the vast, microscopic tapestry of your brain cells and their connections. What we think of as *you* is a vessel of experience into which is poured a small sample of space and time. You imbibe your local culture and technology through your senses. Who you are owes as much to your surroundings as it does to the DNA inside you.

Contrast this story with a Komodo dragon born today and a Komodo dragon born thirty thousand years ago. Presumably it would be more difficult to tell them apart by any measure of their behavior.

What's the difference?

Komodo dragons come to the table with a brain that unpacks to approximately the same outcome each time. The skills on their résumé are mostly hardwired (*eat! mate! swim!*), and these allow them to fill a stable niche in the ecosystem. But they're inflexible workers. If they were airlifted from their home in southeastern Indonesia and relocated to snowy Canada, there would soon be no more Komodo dragons.

In contrast, humans thrive in ecologies around the globe, and soon enough we'll be off the globe. What's the trick? It's not that we're tougher, more robust, or more rugged than other creatures: along any of these measures, we lose to almost every other animal. Instead, it's that we drop into the world with a brain that's largely incomplete. As a result, we have a uniquely long period of helplessness in our infancy. But that cost pays off, because our brains invite the world to shape them—and this is how we thirstily absorb our local languages, cultures, fashions, politics, religions, and moralities.

Dropping into the world with a half-baked brain has proven a winning strategy for humans. We have outcompeted every species on the planet: covering the landmass, conquering the seas, and bounding onto the moon. We have tripled our life spans. We compose symphonies, erect skyscrapers, and measure with ever-increasing precision the details of our own brains. None of those enterprises were genetically encoded.

At least they weren't encoded directly. Instead, our genetics bring about a simple principle: *don't build inflexible hardware; build a system that adapts to the world around it*. Our DNA is not a fixed schematic for building an organism; rather, it sets up a dynamic system that continually rewrites its circuitry to reflect the world around it and to optimize its efficacy within it.

———————

Think about the way a schoolchild will look at a globe of the earth and assume there is something fundamental and unchanging about the country borders. In contrast, a professional historian understands that country borders are functions of happenstance and that our story could have been run with slight variations: a would-be king dies in infancy, or a corn pestilence is avoided, or a warship sinks and a battle tips the other way. Small changes would cascade to yield different maps of the world.

And so it goes with the brain. Although a traditional textbook drawing suggests that neurons in the brain are happily packed next to one another like jelly beans in a jar, don't let the cartoon fool you: neurons are locked in competition for survival. Just like neighboring nations, neurons stake out their territories and chronically defend them. They fight for territory and survival at every level of the system: each neuron and each connection between neurons fights for resources. As the border wars rage through the lifetime of a brain, maps are redrawn in such a way that the experiences and goals of a person are always reflected in the brain's structure. If an accountant drops her career to become a pianist, the neural territory devoted to her fingers will expand; if she becomes a microscopist, her visual cortex will develop higher resolution for the small details she seeks; if she becomes a perfumer, her brain regions assigned to smell will enlarge.

It is only from a dispassionate distance that the brain gives the illusion of a globe with predestined and definitive borders.

The brain distributes its resources according to what's important, and it does so by implementing do-or-die competition among all the parts that make it up. This basic principle will illuminate several questions we'll encounter shortly: Why do you sometimes feel as though your cell phone just buzzed in your pocket, only to discover it's on the table? Why does the Austrian-born actor Arnold Schwarzenegger have a thick accent when he speaks American English, while the Ukrainian-born actress Mila Kunis has none? Why is a child with autistic savant syndrome able to solve a Rubik's Cube in forty-nine seconds but unable to hold a normal conversation with a peer? Can humans leverage technology to build new senses, thus gaining a direct perception of infrared light, global weather patterns, or the stock market?

IF YOU'RE MISSING THE TOOL, CREATE IT

At the end of 1945, Tokyo found itself in a bind. Through the period that spanned the Russo-Japanese War and two world wars, Tokyo had devoted forty years of intellectual resources to military thinking. This had equipped the nation with talents best suited for only one thing: more warfare. But atomic bombs and the fatigue of combat had abated its appetite for conquest in Asia and the Pacific. War was over. The world had changed, and the Japanese nation was going to have to change with it.

But change invited a difficult question: What would they do with their vast numbers of military engineers who, since the dawn of the century, had been trained to produce better weaponry? These engineers simply didn't mesh with Japan's newly discovered desire for tranquility.

Or so it seemed. But over the next few years, Tokyo shifted its social and economic landscape by redeploying its engineers toward new assignments. Thousands were tasked with building the high-speed bullet train known as the Shinkansen.[5] Those who had previously designed aerodynamic navy aircraft now crafted streamlined railcars. Those who had worked on the Mitsubishi Zero fighter plane now devised wheels, axles, and railing to ensure the bullet train could operate safely at high speeds.

Tokyo shaped its resources to better match its external environment. It beat its swords into plowshares. It altered its machinery to match the demands of the present.

Tokyo did what brains do.

The brain chronically adjusts itself to reflect its challenges and goals. It molds its resources to match the requirements of its circumstance. When it doesn't possess what it needs, it sculpts it.

Why is that a good strategy for the brain? After all, human-built technology has been very successful, and we use an entirely different strategy there. We build fixed hardware devices with software programs to neatly accomplish what we need. What would be the advantage of melting the distinction between those layers so that the machinery is constantly redesigned by the running of the programs?

The first advantage is speed.[6] You type rapidly on your laptop because you don't have to think about the details of your fingers' positions, aims, and goals. It all just proceeds on its own, seemingly magically, because typing has become part of your circuitry. By reconfiguring the neural wiring, tasks like this become automatized, allowing fast decisions and actions. Millions of years of evolution didn't presage the arrival of written language, much less a keyboard, and yet our brains have no trouble taking advantage of the innovations. Compare this with hitting the correct keys on a musical instrument you've never played before. For these sorts of untrained tasks, you rely on conscious thinking, and that is comparably quite slow. This speed difference between amateurism and expertise is why a leisure soccer player constantly has the ball stolen. In contrast, the experienced player reads the signals of his opponents, capers with fancy footwork, and shoots the ball with high precision. Unconscious actions are more rapid than conscious deliberation. Plows farm faster than swords.

The second advantage of specializing the machinery for important tasks is energy efficiency. The newbie soccer player simply doesn't understand how all the movement of the field fits together, while the pro can manipulate the game play in multiple ways to score a goal. Whose brain is more active? You might guess it's the high-scoring expert—because he understands the structure of the game and is zipping through possibilities, decisions, and intricate moves. But that would be the wrong guess. The expert's brain has developed neural circuitry specific to soccer, allowing him to make his moves with surprisingly little brain activity. In a sense, the expert has made himself one with the game. In contrast, the amateur's brain is on fire with activity. He's trying to figure out which movements matter. He's entertaining multiple interpretations of the situation and trying to determine which, if any, are correct.

As a result of burning soccer into the circuitry, the pro's performance is both fast and efficient. He's optimized his internal wiring for that which is important in his outside world.

AN EVER-CHANGING SYSTEM

The concept of a system that can be changed by external events—and keep its new shape—led the American psychologist William James to coin the term "plasticity." A plastic object is one that can be shaped, and it can *hold* that shape. This is how the material we call plastic gets its name: we mold bowls, toys, and phones with it, and the material doesn't melt uselessly back to its original form. And so it is with the brain: experience changes it, and it retains the change.

"Brain plasticity" (also called neuroplasticity) is the term we use in neuroscience. But I'll use that term only sparingly in this book, because it sometimes risks missing the target. Whether intentionally or not, "plasticity" suggests that the key idea is to mold something once and keep it that way forever: to shape the plastic toy and never change it again. But that's not what the brain does. It carries on remolding itself throughout your life.

Think of a developing city, and note the way it grows, optimizes, and responds to the world around it. Observe where the city builds its truck stops, how it crafts its immigration policies, how it modifies its education and legal systems. A city is always in flux. A city is not designed by urban planners and then immobilized like a plastic ornament. It incessantly develops.

Just like cities, brains never reach an end point. We spend our lives blossoming toward something, even as the target moves. Consider the feeling of stumbling on a diary entry that you wrote many years ago. It represents the thinking, opinions, and viewpoint of someone who was a bit different from who you are now, and that previous person can sometimes border on the unrecognizable. Despite having the same name and the same early history, in the years between inscription and interpretation the narrator has altered.

The word "plastic" can be stretched to fit this notion of ongoing change, and to keep ties to the existing literature I'll use the term occasionally.[7] But the days of being impressed by plastic molding may be past us. Our goal here is to understand how this living system operates, and for that I'll coin a term that better captures the point: "livewired." As we'll see, it becomes impossible to think about the brain as divisible into layers of hardware and software. Instead, we'll need the concept of liveware to grasp this dynamic, adaptable, information-seeking system.

———

To appreciate the power of a self-configuring organ, let's return to Matthew's story. After the removal of an entire hemisphere of his brain, he was incontinent, couldn't walk, and couldn't speak. His parents' worst fears had materialized.

But with daily physical therapy and language therapy, he was slowly able to relearn language. His acquisition followed the same stages as an infant: first one word, then two, then small phrases.

Three months later, he was developmentally appropriate—right back where he was supposed to be.

Now, many years later, Matthew cannot use his right hand well, and he walks with a slight limp.[8] But he otherwise lives a normal life with little indication that he's been through such an extraordinary adventure. His long-term memory is excellent. He went to college for three semesters, but because of difficulty taking notes with his right hand, he quit to work at a restaurant. There he answers phones, takes care of customer service, serves dishes, and covers just about any job that needs to be done. People who meet him have no suspicion that he is missing half of his brain. As Valerie puts it, "If they didn't know, they wouldn't know."

How could such a major neural obliteration go unnoticed?

Here's how: the remainder of Matthew's brain dynamically rewired to take over the missing functions. The blueprints of his nervous system adjusted themselves to occupy a smaller piece of real estate—encompassing the fullness of life with half the machinery. You couldn't slice out half the electronics from your smartphone and hope to still make a call, because hardware is fragile. Liveware endures.

In 1596, the Flemish cartographer Abraham Ortelius pored over a map of the earth and had a revelation: the Americas and Africa looked as if they could fit together like puzzle pieces. The match seemed clear, but he had no good idea about what had "torn them apart." By 1912, the German geophysicist Alfred Wegener conjectured the notion of continental drift: although the continents had previously been assumed to be immutable in their locations, perhaps they were floating around like mammoth lily pads. The drift is slow (continents waft at the same rate your fingernails grow), but a million-year movie of the globe would reveal the landmasses as part of a dynamic, flowing system, redistributing according to rules of heat and pressure.

Like the globe, the brain is a dynamic, flowing system, but what are its rules? The number of scientific papers on brain plasticity has bloomed into the hundreds of thousands. But even today, as we stare at this strange pink self-configuring material, there is no overarching framework that tells us why and how the brain does what it does. This book lays out that framework, allowing us to better understand who we are, how we came to be, and where we're going.

Once we get in the mode of thinking about livewiring, our current hardwired machines seem hopelessly inadequate for our future. After all, in traditional engineering, everything important is carefully designed. When a car company remodels the chassis of a vehicle, it spends months producing the engine to fit. But imagine changing the bodywork any way you'd like and letting the engine reconfigure itself to match. As we'll see, once we understand the principles of livewiring, we can draft off Mother Nature's genius to fabricate new machines: devices that dynamically determine their own circuitry by optimizing themselves to their inputs and learning from experience.

The thrill of life is not about who we are but about who we are in the process of becoming. Similarly, the magic of our brain lies not in its constituent elements but in the way those elements unceasingly reweave themselves to form a dynamic, electric, living fabric.

Just a handful of pages into this book, your brain has already changed: these symbols on the page have orchestrated millions of tiny changes across the vast seas of your neural connections, crafting you into someone just slightly different than you were at the beginning of the chapter.

2

JUST ADD WORLD

HOW TO GROW A GOOD BRAIN

Brains are not born into the world as blank slates. Instead, they arrive pre-equipped with expectations. Consider the birth of a baby chicken: moments after hatching, it wobbles around on its little legs and can clumsily run and dodge. In its environment, it simply doesn't have time to spend months or years learning how to move around.

Human infants, as well, come to the table with a good deal of pre-programming. Take the fact that we come pre-equipped to absorb language. Or that babies will mimic an adult sticking out her tongue, a feat requiring a sophisticated ability to translate vision into motor action.[1] Or that fibers from your eye don't need to *learn* how to find their targets deep in the brain; they simply follow molecular cues and hit their goal—every time. For all this sort of hardwiring, we can thank our genes.

But genetic hardwiring does not provide the whole story, especially for humans. The system's organization is too complex, and the genes are far too few. Even when you take into account the slicing and dicing that produces many different flavors of the same gene, the number of neurons and their connections vastly outstrips the number of genetic combinations.

So we know that the details of brain wiring involve more than the genetics. And two centuries ago, thinkers began to correctly suspect that the details of experience carried importance. In 1815, the physiologist Johann Spurzheim proposed that the brain, like the muscles, could be increased by exercise: his idea was that blood carried with it the nutrition for growth and that it was "carried in greater abundance to the parts which are excited."[2] By 1874, Charles Darwin wondered if this basic idea might explain why rabbits in the wild had larger brains than domestic rabbits: he suggested that the wild hares were forced to use their wits and senses more than the domesticated ones and that the size of their brains followed.[3]

In the 1960s, researchers began to study in earnest whether the brain could change in measurable ways as a direct result of experience. The simplest way to examine the question was to raise rats in different environments—for example, a rich environment packed with toys and running wheels, or the deprived environment of an empty and solitary cage.[4] The results were striking: the environment altered the rats' brain structure, and the structure correlated with the animals' capacity for learning and memory. The rats raised in enriched environments performed better at tasks and were found at autopsy to have long, lush dendrites (the treelike branches growing from the cell body).[5] In contrast, rats from the deprived environments were poor learners and had abnormally shrunken neurons. This same effect of environment is found in birds, monkeys, and other mammals.[6] To the brain, context matters.

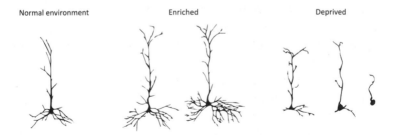

Normal environment   Enriched   Deprived

A neuron normally grows like a branched tree, allowing it to connect to other neurons. In an enriched environment, branches grow more lavishly. In a deprived environment, branches shrivel.

Does the same happen in humans? In the early 1990s, researchers in California realized they could take advantage of autopsies to compare the brains of those who completed high school with those who completed college. In analogy to the animal studies, they found that an area involved in language comprehension contained more elaborate dendrites in the college educated.[7]

So the first lesson is that the fine structure of the brain reflects the environment to which it is exposed. And this is not just about dendrites. As we'll learn shortly, world experience modulates almost every measurable detail of the brain, from the molecular scale to overall brain anatomy.

EXPERIENCE NECESSARY

Why was Einstein *Einstein*? Surely genetics mattered, but he is affixed to our history books because of every experience he'd had: the exposure to cellos, the physics teacher he had in his senior year, the rejection of a girl he loved, the patent office in which he worked, the math problems he was praised for, the stories he read, and millions of further experiences—all of which shaped his nervous system into the biological machinery we distinguish as Albert Einstein. Each year, there are thousands of other children with his potential but who are exposed to cultures, economic conditions, or family structures that don't give sufficiently positive feedback. And we don't call them Einsteins.

If DNA were the only thing that mattered, there would be no particular reason to build meaningful social programs to pour good experiences into children and protect them from bad experiences. But brains require the right kind of environment if they are to correctly develop. When the first draft of the Human Genome Project came to completion at the turn of the millennium, one of the great surprises was that humans have only about twenty thousand genes.[8] This number came as a surprise to biologists: given the complexity of the brain and the body, it had been assumed that hundreds of thousands of genes would be required.

So how does the massively complicated brain, with its eighty-six billion neurons, get built from such a small recipe book? The answer pivots on a clever strategy implemented by the genome: build incompletely and let world experience refine. Thus, for humans at birth, the brain is remarkably unfinished, and interaction with the world is necessary to complete it.

Consider the sleep-wake cycle. This internal clock, known as the circadian rhythm, runs roughly on a twenty-four-hour cycle. However, if you descend into a cave for several days—where there are no clues to the light and dark cycles of the surface—your circadian rhythm would drift in a range between twenty-one and twenty-seven hours. This exposes the brain's simple solution: build a non-exact clock and then calibrate it to the sun's cycle. With this elegant trick, there is no need to genetically code a perfectly wound clock. The world does the winding.

The flexibility of the brain allows the events in your life to stitch themselves directly into the neural fabric. It's a great trick on the part of Mother Nature, allowing the brain to learn languages, ride bicycles, and grasp quantum physics, all from the seeds of a small collection of genes. Our DNA is not a blueprint; it is merely the first domino that kicks off the show.

From this viewpoint, it is easy to understand why some of the most common problems of vision—such as the inability to see depth correctly—develop from imbalances in the pattern of activity delivered to the visual cortex by the two eyes. For example, when children are born cross-eyed or wall-eyed, the activity from the two eyes is not well correlated (as it would be with aligned eyes). If the problem is not addressed, the child will not develop normal stereo vision—that is, the capacity to determine depth from the small differences between what the two eyes are seeing. One eye will grow progressively weaker, often to the point of blindness. We'll return to this later to understand why and what can be done about it. For now, the important point is that the development of normal visual circuits relies on normal visual input. It is *experience-dependent*.

So genetic instructions play only a minor role in the detailed assembly of cortical connections. It couldn't be any other way: With twenty thousand genes and 200 trillion connections between neurons, how could the details possibly be prespecified? That model could never have worked. Instead, neuronal networks require interaction with the world for their proper development.[9]

NATURE'S GREAT GAMBLE

On September 29, 1812, a baby was born who would inherit the grand ducal throne of Baden, Germany. Unfortunately, the baby died seventeen days later. That was the end of that.

Or was it? Sixteen years later, a young man named Kaspar Hauser showed up in Nuremberg, Germany. He carried a note that explained he had been given away as a child, and he apparently knew only a few sentences, including "I want to be a cavalryman, as my father was." He attracted widespread attention and the audiences of powerful people; many began to suspect he was the hereditary prince of Baden, switched in those first weeks with a dying baby in a nefarious plot by those who stood to inherit the throne.

The story grew famous beyond the royal intrigue: Kaspar became the exemplar of a feral child. According to his own telling, Kaspar had spent his entire youth alone in a dark cell. It was only a meter wide, two meters long, and one and a half meters high. It had a straw bed and a small wooden horse. Each morning he awoke to discover some bread and water, nothing more. He saw no one enter or leave. Occasionally the water he drank would taste a bit different, and then he would grow sleepy afterward—and when he awoke his hair would be cut and his nails clipped. But it was not until just before his release that he had direct contact with another human, a man who taught him how to write but always kept his face hidden from view.

Kaspar Hauser's story aroused international attention. He grew up to write prolifically and touchingly about his childhood. His story lives on today in plays, books, and music; it is perhaps history's most famous story of a feral childhood.

But Kaspar's claim was almost certainly false. Beyond the extensive historical analysis that rules it out, there is a neurobiological reason: a child raised without human interaction does not grow up to walk, speak, write, lecture, and thrive, like the successful Kaspar. After a century of popular press about Kaspar, the psychiatrist Karl Leonhard put a fine point on it:

> If he had been living since childhood under the conditions he describes, he would not have developed beyond the condition of an idiot; indeed he would not have remained alive long. His tale is so full of absurdities that it is astonishing that it was ever believed and is even today still believed by many people.[10]

After all, despite some genetic pre-specification, nature's approach to growing a brain relies on receiving a vast set of experiences, such as social interaction, conversation, play, exposure to the world, and the rest of the landscape of normal human affairs. The strategy of interaction with the world allows the colossal machinery of the brain to take shape from a relatively small set of instructions. It's an ingenious approach for unpacking a brain (and body) from a single microscopic egg.

But this strategy is also a gamble. It's a slightly risky approach—one in which the brain-shaping work is partially relegated to world experience rather than hardwiring. After all, what if a child is actually born into Kaspar's story and has an infancy characterized by total parental neglect?

Tragically, we know the answer to this question. In one example in July 2005, police in Plant City, Florida, pulled up outside a dilapidated house to perform an investigation. They had been alerted by a neighbor who had seen a girl in the window on a few occasions but had never seen the girl exit the house and had never seen any adults with her in the window.

The officers knocked on the door for a while and were eventually greeted by a woman. They told the woman they had a warrant to search for her daughter inside the house. They walked down the hallways, probed several rooms, and finally entered a small bedroom. There was the girl. One of the officers vomited.

Danielle, a feral child discovered in 2005 in Florida.
Although the photograph displays a child's beautiful face, the behaviors
and expressions inherent to normal human interaction are absent in her:
she missed the critical window for proper input from the world.

Danielle Crockett, an undersized girl of almost seven years old, had been locked away in a dark closet for her entire childhood. She was flecked with fecal matter and cockroaches. Beyond basic sustenance, she had never received physical affection, never engaged in normal conversation, and in all likelihood had never been let outdoors. She was fully incapable of speech. When she met the police officers (and later the social workers and psychologists), she appeared to look right through them; she had no glimmer of recognition or indications of normal human interaction. She could not chew solid food, did not know how to use a toilet, could not nod yes or no, and one year later had not mastered the use of a sippy cup. After many tests, physicians were able to verify that she had no genetic problems such as cerebral palsy, autism, or Down syndrome. Instead, the normal development of her brain had been derailed by severe social deprivation.

Despite the best attempts of doctors and social workers, the prognosis for Danielle is poor; the likely scenario is that she will live in a nursing home and may eventually be able to live without diapers.[11] Heartbreakingly, hers is a real-life Kaspar Hauser story, with the real-life consequences.

Danielle's outcome is grim because the human brain arrives in the world unfinished. Proper development requires proper input. The brain absorbs experience to unpack its programs, and only during a rapidly closing window of time. Once the window is missed, it is difficult or impossible to reopen.

Danielle's story is paralleled by a set of animal experiments in the early 1970s. Harry Harlow, a scientist at the University of Wisconsin, used monkeys to study the bonding between mothers and their children. He had an active scientific career, but when his wife died of cancer in 1971, Harlow sank into a depression. He continued to work, but his friends and co-workers sensed that he was not the same. He turned his scientific interests to the study of depression.

Using monkeys to model human depression, he developed a study of isolation. He put a baby monkey into a steel-walled cage with no windows. A two-way mirror allowed Harlow to look in but prevented the monkey from seeing out. Harlow tried this with a monkey for thirty days. Then another monkey for six months. Other monkeys were incarcerated for a full year.

Because the baby monkeys never had the chance to develop normal bonds (they were put in the cage shortly after birth), they emerged with deep-seated disturbances. Those who were isolated the longest ended up much like Danielle: they showed no normal interaction with other monkeys and did not engage in recreation, cooperation, or competition. They barely moved. Two would no longer eat.

Harlow also noted that the monkeys were incapable of having normal sexual relations. Even so, he took some of the isolated females and had them impregnated to see how these disturbed monkeys would interact with children of their own. The results were disastrous. The isolated monkeys were completely unable to raise children. In the best cases, they ignored the children entirely; in the worst cases, they injured them.[12]

The lesson of Harlow's monkeys is the same as the lesson from Danielle: Mother Nature's strategy of unpacking a brain relies on proper world experience. Without it, the brain becomes malformed and pathological. Like a tree that needs nutrient-rich soil to arborize, a brain requires the rich soil of social and sensory interaction.

———————

With this background in hand, we now see that the brain leverages its environment to shape itself. But how, exactly, does it absorb the world—especially from inside its dark cave? What happens when a person loses an arm or goes deaf? Does a blind person actually enjoy better hearing? And what does any of this have to do with why we dream?

3

THE INSIDE MIRRORS THE OUTSIDE

THE CASE OF THE SILVER SPRING MONKEYS

In 1951, the neurosurgeon Wilder Penfield sank the tip of a fine electrode into the brain of a man undergoing surgery.[1] Along the brain tissue just beneath where one might wear headphones, Penfield discovered something surprising. If he gave a small shock of electricity at a particular spot, the patient would feel as though his hand were being touched. If Penfield stimulated a nearby spot, the patient would feel the touch on his torso. A different spot, the knee. Every spot on the patient's body was represented in the brain.

Then Penfield made a deeper realization: neighboring parts of the body were represented by neighboring spots on the brain. The hand was represented near the forearm, which was represented near the elbow, which was represented near the upper arm, and so on. Along this strip of the brain, there was a detailed map of the body. By moving from spot to spot slowly along the somatosensory cortex, he could find the whole human figure.[2]

And this wasn't the only map he found. Along the motor cortex (the strip just in front of the somatosensory cortex), he discovered the same kind of result: a little zap of electricity caused muscles to twitch in specific, neighboring areas of the body. Again, it was laid out in an orderly manner.

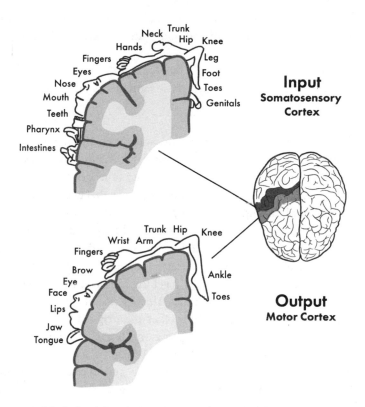

Maps of the body are found where inputs enter the brain (somatosensory cortex, top) and outputs leave the brain (motor cortex, bottom). Areas with more detailed sensation, and those that are more finely controlled, command more real estate.

He named these maps of the body the homunculus, or "little man."

But the existence of the maps is strange and unexpected. How do they *exist*? After all, the brain is locked in total darkness within the skull. These three pounds of tissue don't know what your body looks like; the brain has no way to directly *see* your body. It has access to nothing but a chattering stream of electrical pulses racing up the thick bundles of data cables we call nerves. Stowed away in its bony prison, the brain should have no idea what limbs are connected where, or which are next to which others. So how is there a depiction of the body's layout in this lightless vault?

A moment of thought will likely lead you to the most straightforward solution: the map of the body must be genetically preprogrammed. Good guess!

But wrong.

Instead, the answer to the mystery is more fiendishly clever.

––––––––––

A clue to the map mystery came decades later, in an unexpected turn of events. Edward Taub, a scientist at the Institute of Behavioral Research in Silver Spring, Maryland, wanted to understand how victims of brain injury could recover movement. To that end, he obtained seventeen monkeys and studied whether severed nerves could regenerate. In each, he carefully cut a nerve bundle that linked the brain to one of the arms or one of the legs. As expected, the unfortunate monkeys lost all sensation from the affected limbs, and Taub set about studying whether there was a way to get the monkeys to regain use.

In 1981, a young volunteer named Alex Pacheco began to work in the lab. Although he presented himself as an intrigued student, in fact he was there to spy for a budding organization, the People for the Ethical Treatment of Animals (PETA). At night Pacheco took photographs. Some of the shots were apparently staged to exaggerate the suffering of the monkeys,[3] but in any case the effect was achieved. In September 1981, the Montgomery County police raided the laboratory and shut it down. Dr. Taub was convicted of six counts of failure to provide adequate veterinary care. All of the charges were overturned on appeal; nonetheless, the events led to the creation of the Animal Welfare Act of 1985, in which Congress defined new rules for animal care in research environments.

Although this provided a watershed moment for animal rights, the importance of the story is not only about what happened in Congress. For our purposes here, it's about what happened to the seventeen monkeys. Immediately after the accusation, PETA swept in and absconded with the monkeys, leading to charges of theft of court evidence. Incensed, Taub's research institution demanded the return of the monkeys. The legal battle grew increasingly heated, and the battle for possession of the monkeys ascended to the U.S. Supreme Court.

The Court rejected PETA's plea to keep the monkeys, instead granting custody to a third party, the National Institutes of Health. While humans barked at each other in distant courtrooms, the disabled monkeys enjoyed an early retirement by eating, drinking, and playing together for ten years.

Near the end of this period, one of the monkeys became terminally ill. The court agreed that the monkey could be put to sleep. And here's where the plot turned. A group of neuroscience researchers made a proposal to the judge: the monkey's severed nerve would not have been in vain if the researchers could be allowed to perform a brain-mapping study on the monkey while it was under anesthesia, just before being euthanized. After some debate, the court granted permission.

On January 14, 1990, the research team put recording electrodes in the monkey's somatosensory cortex. Exactly as Wilder Penfield had done with his human patient, the researchers touched the monkey on its hand, arm, face, and so on while recording from neurons in the brain. In this way, they revealed the map of the body in the brain.

The findings sent ripples through the neuroscience community. The body map had changed over the years. Unsurprisingly, a gentle touch on the monkey's nerve-severed hand no longer activated any response in the cortex. But the surprise was that the little bit of cortex that used to represent the hand was now excited by a touch to the face.[4] The map of the body had reorganized. The homunculus still looked like a monkey, but a monkey without a right arm.

This discovery ruled out the possibility that the brain's map of the body is genetically preprogrammed. Instead, something much more interesting was going on. The brain's map was flexibly defined by active inputs from the body. When the body changes, the homunculus follows.

The same brain-mapping studies were done later in the year on the other Silver Spring monkeys. In each one of them, the somatosensory cortex had dramatically rearranged: the areas once representing the nerve-severed limbs had been taken over by neighboring areas in the cortex. The homunculi had transformed to match the monkeys' new body plans.[5]

What does it *feel* like when the brain reorganizes like that? Unfortunately, monkeys can't tell us. But people can.

THE AFTERLIFE OF LORD HORATIO NELSON'S RIGHT ARM

The British naval commander Admiral Lord Horatio Nelson (1758–1805) is the hero mounted high on a pedestal overlooking London's Trafalgar Square.[6] The statue stands in towering testimony to his charismatic leadership, his tactical strength, and his inventive stratagems, which together led to decisive victories on waters from the Americas to the Nile to Copenhagen. He died heroically in his final showdown—the Battle of Trafalgar—one of Britain's greatest maritime victories.

Beyond his naval impact, Admiral Nelson also contributed to neuroscience—however, this was totally accidental. His involvement began during his attack on Santa Cruz de Tenerife, when at eleven o'clock at night on July 24, 1797, a musket ball exited a Spanish rifle barrel at a thousand feet per second and ended its trajectory in Lord Nelson's right arm. His bone shattered. Nelson's stepson tied a piece of his neck scarf tightly around the arm to stop the bleeding, and Nelson's sailors rowed vigorously back to the main ship, where the surgeon tensely awaited. After a rapid physical exam, the good news was that Nelson was likely to survive. The bad news was that the risk of gangrene demanded amputation. Nelson's right arm was surgically removed above the elbow and thrown overboard into the water.

Over the following weeks, Nelson learned to function without his right arm—eating, washing himself, even shooting. He came to jokingly refer to the stump of his amputation as his "fin."

But some months after the event, strange consequences began. Lord Nelson started to feel—literally *feel*—that his arm was still present. He experienced sensations from it. He was certain that his missing fingernails from his missing fingers were digging, painfully, into his missing right palm.

Nelson had an optimistic interpretation of this sensation of his phantom limb: he concluded that he now possessed incontrovertible

Although paintings and sculptures of Lord Horatio Nelson festoon British museums, most visitors don't notice that Nelson is missing his right arm. Its amputation in 1797 led to an early clinical case of phantom limb sensation and an interesting, but incorrect, metaphysical interpretation by Nelson himself.

proof of life after death. After all, if an absent limb could give rise to conscious feeling—an ever-present ghost of itself—then an absent body must as well.

Nelson was not the only one to notice these strange sensations. Across the Atlantic some years later, a physician named Silas Weir Mitchell documented numerous Civil War amputees at a hospital in Philadelphia. He was mesmerized by the fact that many of them insisted they still felt sensations from their missing limbs.[7] Was this proof of Nelson's corporeal immortality?

As it turns out, Nelson's conclusion was premature. His brain was remapping itself, exactly as happened with the Silver Spring monkeys. Over time, as historians followed the shifting borders of the

British Empire, scientists discovered how to track the shifting borders in the human brain.[8] With modern imaging techniques, we can see that when an arm is amputated, its representation in the cortex is encroached upon by neighboring areas. In this case, the cortical areas that surround the hand and forearm are the territories of the upper arm and the face. (Why the face? It just happens that's where things lie when the body has to be represented on a linear map.) So these representations move to take over the land where the hand used to be. Just as with the monkeys, the maps come to reflect the current form of the body.

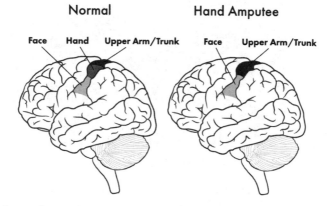

The brain adapts to the body plan. When a hand is amputated, neighboring cortical territories move in to usurp the hand's previously held territory.

However, there's another mystery buried in here. Why did Nelson still have a sense of his hand, and why, if you were to touch Nelson on his face, would he say that his phantom hand was being touched? Didn't the neighboring areas take over the hand representation? The answer is that touch to the hand is represented not only by cells in the somatosensory cortex but also by the cells they talk to downstream, and the cells *they* talk to. So although the map modified itself rapidly in the primary somatosensory cortex, it shifted less and less in downstream areas. In a child born without an arm, the map would be entirely different—but in an adult, like Lord Nelson, the system has less flexibility to rewrite its manifest. Deep in Lord Nelson's brain, the neurons downstream of the somatosensory cortex did not shift their connections as much, and therefore they believed that any activity they received was due to touch on the hand. As a result, Nelson perceived the ghostly presence of his missing limb.[9]

———————

Monkeys and admirals and civil war veterans tell the same story: when inputs suddenly cease, sensory cortical areas do not lie fallow. Instead, they are invaded by their neighbors.[10] With thousands of amputees now studied in brain scanners, we see the degree to which brain matter is not like hardware, but instead dynamically reallocates.

Although amputations lead to dramatic cortical reorganization, the brain's shape shifting can be induced by modifying the body in more modest ways. For example, if I were to fasten a tight pressure cuff to your arm, your brain would adjust to the weakened incoming signals by devoting less territory to that part of your body.[11] The same thing happens if the nerves from your arm are blocked for a long time with anesthetics. In fact, if you merely tie two fingers of your hand together—so they no longer operate independently, but instead as a unit—their cortical representation will eventually merge from two distinct regions into a single area.[12]

So how does the brain, confined to its dark perch, keep constant track of what the body looks like?

TIMING IS EVERYTHING

Imagine taking a bird's-eye view of your neighborhood. You notice that some people take their dogs for a walk every morning at six o'clock. Others don't get out with their canines until nine. Others stroll their pooches after lunch. Others opt for nighttime walks. If you watched the dynamics of the neighborhood for a while, you'd notice that people in the neighborhood who happen to walk at the same time tend to become friends with one another: they bump into one another, they chat, they eventually invite each other over for barbecues. Friendship follows timing.

It's the same with neurons. They spend a small fraction of their time sending abrupt electrical pulses (also called spikes). The timing of these pulses is critically important. Let's zoom in to a typical neuron. It reaches out to touch ten thousand neighbors. But it doesn't form equally strong relationships with all ten thousand. Instead, the strengths are based on timing. If our neuron spikes, and then a connected neuron spikes just after that, the bond between them is strengthened. This rule can be summarized as *neurons that fire together, wire together.*[13]

In the young neighborhood of a new brain, nerves coming from the body to the brain branch out broadly. But they set down permanent roots in places where they fire in close timing with other neurons. Because of the synchrony, they strengthen their bonds. They don't host barbecues, but instead they release more neurotransmitters, or set up more receptors to receive the neurotransmitters, thus causing a stronger link between them.

How does this simple trick lead to a map of the body? Consider what happens as you bump, touch, hug, kick, hit, and pat things in the world. When you pick up a coffee mug, patches of skin on your fingers will tend to be active at the same time. When you wear a shoe, patches of skin on your foot will tend to be active at the same time. In contrast, touches on your ring finger and your little toe will tend to enjoy less correlation, because there are few situations in life when those are active at the same moment. The same is true all over your body: patches that are neighboring will tend to be co-active more than patches that are not neighboring. After interacting with the world for a while, areas of skin that happen to be co-active often will wire up next to one another, and those that are not correlated will tend to be far apart. The consequence of years of these co-activations is an atlas of neighboring areas: a map of the body. In other words, the brain contains a map of the body because of a simple rule that governs how individual brain cells make connections with one another: neurons that are active close in time to one another tend to make and maintain connections between themselves. That's how a map of the body emerges in the darkness.[14]

But why does the map change when the input changes?

COLONIZATION IS A FULL-TIME BUSINESS

At the beginning of the seventeenth century, France began its colonization of North America. Its technique? Sending ships full of Frenchmen. It worked. The French settlers took root in the fresh territory. In 1609, the French erected a fur post that would eventually become the city of Quebec, which was destined to become the capital of New France. Within twenty-five years, the French had spread into Wisconsin. As new French settlers voyaged across the Atlantic, their territory grew.

But New France wasn't easy to maintain: it was under constant competition from the other powers that were sailing ships that way, mostly Britain and Spain. So France's king, Louis XIV, started to intuit an important lesson: if he wanted New France to firmly take root, he had to keep sending ships—because the British were sending even *more*

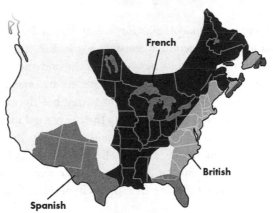

North America, 1750

ships. He understood that Quebec wasn't growing rapidly enough because of a lack of women, and so he sent 850 young women (called King's Daughters) to stimulate the local French population. The effort helped to lift the population of New France to seven thousand by 1674 and then to fifteen thousand by 1689.

The problem was that the British were sending far more young men and women. By 1750, when New France had sixty thousand inhabitants, Britain's colonies boasted a million. That made all the difference in the subsequent wars between the two powers: despite their allegiances with the Native Americans, the French were badly outstripped. For a short time, the government of France forced newly released prisoners to marry local prostitutes, and then the newlywed couples were linked with chains and shipped off to Louisiana to settle the land. But even these French efforts were insufficient.

By the end of their sixth war, the French realized they had lost. New France was dissolved. The spoils of Canada moved under the control of Great Britain, and the Louisiana Territory went to the young United States.[15]

The waxing and waning of the French grip on the New World had everything to do with how many boats were being sent over. In the face of fierce competition, the French had simply not shipped enough people over the water to keep a hold on their territory. As a result, all that now remains of the French presence in the New World are linguistic fossils, as seen in place-names such as Louisiana, Vermont, and Illinois.

Without competition, colonization is easy, but in the face of rivalry holding on to territory requires constant work. The same story plays out constantly in the brain. When a part of the body no longer sends information, it loses territory. Admiral Nelson's arm was France, and his cortex the New World. It started off with a healthy colonization, sending useful spikes of information up the nerves and into the brain, and in Nelson's youth it staked out a healthy territory. But then came the musket ball, followed hours later by his tattered arm splashing into the dark water . . . and now his brain received no new input from that part of his body. With time, the arm lost its neural real estate. Eventually, all that remained were fossils of the arm's former presence, such as a feeling of phantom pain.

These lessons of colonization apply to more than arms: they apply to any system sending information into the brain. When a person's eyes are damaged, signals no longer flood in along the pathways to the occipital cortex (the portion at the back of the brain, often thought of as "visual" cortex). And so that part of the cortex becomes no longer visual. The ships carrying visual data have stopped arriving, so the coveted territory is taken over by the competing kingdoms of sensory information.[16] As a result, when a blind person passes her fingertips over the raised dots of a Braille poem, her occipital cortex becomes active from mere touch.[17] If she gets a stroke that damages her occipital cortex, she'll lose her ability to understand Braille.[18] Her occipital cortex has been colonized by touch.

And it's not only touch, but any sources of information. When blind

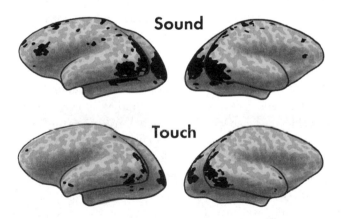

Cortical reorganization: unused cortex is taken over by competing neighborhoods.
In this brain scan, sound and touch activate the otherwise unused occipital
cortex of the blind (black indicates regions more active in the blind than the sighted).
For a better view of the hills and valleys of the cortex, the brain has been
computationally "inflated." Figure adapted from Renier et al. (2010).

subjects listen to sounds, their auditory cortex becomes active, and so does their occipital cortex.[19]

Not only can touch and sound activate the previously visual cortex of the blind, but so can smell, taste, the reminiscence of events, or the solving of math problems.[20] As with a map of the New World, territory goes to the fiercest competitors.

The story has grown even more interesting in recent years: when new occupants move into the visual cortex, they retain some of the former architecture—like the mosques in Turkey that used to be Roman cathedrals. As an example, the area that processes visual written language in the sighted is the same area that becomes active when the blind read Braille.[21] Similarly, the main area for processing visual motion in the sighted is activated for tactile motion in the blind (for example, something moving across the fingertips or the tongue).[22] The main neural network involved in visual object recognition in the sighted is activated by touch in the blind.[23] Such observations have led to the hypothesis that the brain is a "task machine"—doing jobs like detecting motion or objects in the world—rather than a system organized by particular senses.[24] In other words, brain regions care about solving certain types of tasks, irrespective of the sensory channel by which information arrives.

There's a side note here that we'll return to in later chapters: age matters. In those born blind, their occipital cortex is completely taken over by other senses. If a person goes blind at an early age—say, at five years old—the takeover is less comprehensive. For the "late blind" (those who lost vision after the age of ten), the cortical takeovers are even smaller. The older the brain, the less flexible it is for redeployment, just as North American borders now shift very little after settling into place for five centuries.

The same thing we see with the loss of vision happens with the loss of any sense. For example, in the deaf, the auditory cortex becomes employed for vision and other tasks.[25] Just as Lord Nelson's loss of a limb led to cortical takeovers by neighboring territories, so too does the loss of hearing, smell, taste, or anything else. The cartography of the brain constantly shifts to best represent the incoming data.[26]

Once you begin looking for it, you'll see this competition for territory everywhere around you. Think of an airport in a major city. If there are a large number of incoming flights from a particular airline (United), and fewer flights from another (Delta), then it would be no surprise to see the number of United counters grow, while those from Delta shrink. United would take over more of the gates, more of the baggage claims, and more space on the monitors. If another airline went fully out of business (think Trans World Airlines), then all of its presence in the airport would be quickly taken over. And so it goes with the brain and its sensory inputs.

Now we understand how competition leads to takeover. But all this leads to a question: When a sense captures more area, does that give it greater capabilities?

THE MORE THE BETTER

A young boy named Ronnie was born in Robbinsville, North Carolina. Soon after he was born, it became clear he was blind. At one year and one day old, his mother abandoned him, citing that his blindness was her punishment from God. He was raised in poverty by his grandparents until he was five, then sent off to a school for the sightless.

When he was six years old, his mother came by, just once. She had another child now, a girl. His mother said, "Ron, I want you to feel her eyes. You know, her eyes are so pretty. She did not shame me the way that you did. She can see." That was the last time he ever had contact with his mother.

As hard as his childhood was, it became clear that Ronnie had a gift for music. His instructors spotted his talents, and Ronnie began to formally study classical music. One year after he picked up the violin, his teachers declared him a virtuoso. He went on to master piano, guitar, and several other string and woodwind instruments.

From there he went on to become one of the most popular performers of his day, locking down both pop music and country-western markets. He secured forty country hits at the number-one slot. He earned six Grammy Awards.

Ronnie Milsap is just one of many blind musicians; others include Andrea Bocelli, Ray Charles, Stevie Wonder, Diane Schuur, José Feliciano, and Jeff Healey. Their brains have learned to rely on the signals of sound and touch in their environment, and they become better at processing those than sighted people.

While musical stardom is not guaranteed for blind people, brain reorganization is. As a result, perfect musical pitch is overrepresented in the blind, and blind people are up to ten times better at determining whether a musical pitch subtly wobbles up or down.[27] They simply have more brain territory devoted to the task of listening. In a recent experiment, participants who were sighted or blind had one ear plugged up and were then asked to point to the locations of sounds in the room. Because pinpointing a sound requires a comparison of the signals at both ears, it was expected that everyone would fail miserably at this task. And that's what happened with the sighted participants. But the blind participants were able to generally tell where the sounds were positioned.[28] Why? Because the exact shape of the cartilage of the outer ear (even just one ear) bounces sounds around in subtle ways that give clues to location—but only if one is highly attuned to pick up on those signals. The sighted have less cortex devoted to sound, and so their ability to extract subtle sound information is underdeveloped.

This sort of extreme talent with sound is common among the blind. Take Ben Underwood. When he was two years old, Ben stopped seeing out of his left eye. His mother took him to the doctor and soon discovered he had retinal cancer in both eyes. After chemotherapy and radiation failed, surgeons removed both his eyes. But by the time Ben was seven years old, he had devised a useful, unexpected technique: he clicked with his mouth and listened for the returning echoes. By this method, he could determine the locations of open doorways, people, parked cars, garbage cans, and so on. He was echolocating: bouncing his sound waves off objects in the environment and listening to what returned.[29]

A documentary about Ben kicked off with the statement that he was "the only person in the world who can see with echolocation."[30] The statement was erroneous in a couple of ways. First, Ben may or may not have *seen* the way a sighted person would think of sight; all we know is that his brain was able to convert sound waves into some practical understanding of the large objects in front of him. But more on that later.

Second, and more important, Ben was not the only one using echolocation: thousands of blind people have done so.[31] In fact, the phenomenon has been discussed since at least the 1940s, when the word "echolocation" was first coined in a *Science* article titled "Echolocation by Blind Men, Bats, and Radar."[32] The author wrote, "Many blind persons develop in the course of time a considerable ability to avoid obstacles by means of auditory cues received from sounds of their own making." This included their own footsteps, or cane tapping, or finger snapping. He demonstrated that their ability to successfully echolocate was drastically reduced by distracting noises or earplugs.

As we saw earlier, the occipital lobe can be taken over by many tasks, not just those of hearing. Memorization, for example, can benefit from the extra cortical real estate. In one study, blind people were tested to see how well they could remember lists of words. Those with *more* of their occipital cortex taken over could score higher: they had more territory to devote to the memory task.[33]

The general story is straightforward: the more real estate, the better. This sometimes leads to counterintuitive results. Most people are born with three different types of photoreceptors for color vision, but some people are born with only two types, one type, or none, giving them diminished (or no) ability to discriminate among colors. However, color-blind people don't have it all bad: they are *better* at distinguishing between shades of gray.[34] Why? Because they have the same amount of visual cortex, but fewer color dimensions to worry about. Using the same amount of cortical territory available for a simpler task gives improved performance. Although the military excludes color-blind soldiers from certain jobs, they have come to realize that the color-blind can spot enemy camouflage better than people with normal color vision.

And although we've been using the visual system to introduce the critical points, cortical redeployment happens everywhere. When people lose hearing, previously "auditory" brain tissue comes to represent other senses.[35] Thus, it won't come as a surprise that deaf people have better peripheral visual attention, or that they can often *see* your accent: they can tell what part of the country you're from, because they're so good at lip-reading. Similarly, after an amputee loses a limb, the sensation on the stump becomes finer. Touch can now be sensed with lighter pressure, and two touches close together can be sensed as separate touches rather than a single one. Because the brain is now devoting more territory to the remaining, undamaged areas, sensing becomes higher resolution.

Neural redeployment replaces the old paradigm of predetermined brain areas with something more flexible. Territory can be reassigned to different tasks. There is nothing special about *visual* cortex neurons, for example. They are simply neurons that happen to be involved in processing edges or colors in people who have functioning eyes. These exact same neurons can process other types of information in the sightless.

The old paradigm would assert that the North American acreage labeled as Louisiana was predetermined for French people. The new paradigm is not surprised when the Louisiana Territory gets sold and citizens from around the globe set up shop there.

Given that the brain has to distribute all its tasks across the finite volume of the cortex, it may be that some disorders arise from suboptimal distributions. One example is autistic savantism, in which a child who has severe cognitive and social deficits might be a virtuoso at, say, memorizing the phone book, or copying down visual scenes, or solving the Rubik's Cube with stunning speed. The pairing of cognitive disabilities with outstanding talents has attracted many theories; one of relevance here is an unusual distribution of cortical real estate.[36] The idea is that atypical feats can be accomplished when the brain devotes an unusually large swath of its real estate to one task (such as memorization, or visual analysis, or puzzles). But these human superpowers come at the expense of other tasks among which brains normally divide their territory, such as all the subtasks that add up to reliable social skills.

BLINDINGLY FAST

Recent decades have yielded several revelations about brain plasticity, but perhaps the biggest surprise is its rapidity. Some years ago, researchers at McGill University put several adults who had just recently lost their sight into a brain scanner. The participants were asked to listen to sounds. Not surprisingly, the sounds caused activity in their auditory cortex. But the sounds also caused activity in their occipital cortex—activity that would not have been there even a few weeks earlier, when the participants had sight. The activity wasn't as strong as that seen in people who had been blind for a long while, but it was detectable nevertheless.[37]

This demonstrated that the brain can implement changes rapidly when vision disappears. But how rapidly?

The researcher Alvaro Pascual-Leone began to wonder about the speed at which these major brain changes can take place. He noted that aspiring instructors at a school for the blind were required to blindfold themselves for seven full days to gain firsthand understanding of their students' living experiences. Most of the instructors become aware of enhanced skills with sounds—orienting to them, judging their distance, and identifying them:

> Several describe becoming able to identify people quickly and accurately as they started talking or even as they simply walked by due to the cadence of their steps. Several learned to differentiate cars by the sounds of their motors, and one described the "joy of telling motorcycles apart by their sound."[38]

This got Pascual-Leone and his colleagues considering what would happen if a sighted person were blindfolded in a laboratory setting for several days. They launched the experiment, and what they found was nothing short of remarkable. They discovered that neural reorganization—the same kind seen in blind subjects—also happens with temporary blindness of sighted subjects. Rapidly.

In one of their studies, sighted participants were blindfolded for five days, during which time they were put through an intensive Braille-training paradigm.[39] At the end of five days, the subjects had become quite good at detecting subtle differences between Braille characters—much better than a control group of sighted participants who underwent the same training without a blindfold.

But especially striking was what happened to their brains, as measured in the scanner. Within five days, the blindfolded participants had recruited their occipital cortex when they were touching objects. Control subjects, not surprisingly, used only their somatosensory cortex. The blindfolded subjects also showed occipital responses to sounds and words.

When this new occipital lobe activity was intentionally disrupted in the laboratory by magnetic pulses, the Braille-reading advantage of the blindfolded subjects went away—indicating that the recruitment of this brain area was not an accidental side effect but a critical piece of the improved behavioral performance.

When the blindfold was removed, the response of the occipital cortex to touch or sound disappeared within a day. At that point, the participants' brains returned to looking indistinguishable from every other sighted brain out there.

In another study, the visual areas of the brain were carefully mapped out using more powerful neuroimaging techniques. Participants were blindfolded, put in a scanner, and asked to perform a touching task that required fine discrimination with their fingers. In these conditions, investigators could detect activity emerging in the primary visual cortex after a blindfolding session of a mere forty to sixty minutes.[40]

The shock of these findings was their sheer speed. The shape shifting of brains is not like the glacial drifting of continental plates, but can instead be remarkably swift. In later chapters, we'll see that visual deprivation causes the unmasking of *already-existing* nonvisual input into the occipital cortex, and we'll come to understand how the brain is always sprung like a mousetrap to implement rapid change. But for now the important point is that the brain's changes are more brisk than even the most optimistic neuroscientist would have dared to guess at the beginning of this century.

Let's zoom back out to the bigger picture. Just as sharp teeth and fast legs are useful for survival, so is neural flexibility: it allows brains to optimize performance in a variety of environments.

But the competition in the brain has a potential downside as well. Whenever there's an imbalance of activity in the senses, a potential takeover can happen, and it can happen rapidly. A redistribution of resources can be optimal when a limb or a sense has been permanently amputated or lost, but the rapid conquest of territory may have to be actively combated in other scenarios. And this consideration led me and my former student Don Vaughn to propose a new theory for what happens to brains in the dark of night.

WHAT DOES DREAMING HAVE TO DO WITH THE ROTATION OF THE PLANET?

One of the unsolved mysteries in neuroscience is why brains dream. What are these bizarre nighttime hallucinations about? Do they have meaning? Or are they simply random neural activity in search of a coherent narrative? And why are dreams so richly visual, igniting the occipital cortex every night into a conflagration of activity?

Consider the following: In the chronic and unforgiving competition for brain real estate, the visual system has a unique problem to deal with. Because of the rotation of the planet, it is cast into darkness for an average of twelve hours every cycle. (This refers to 99.9999 percent of our species' evolutionary history, not to the current, electricity-blessed times.) We've already seen that sensory deprivation triggers neighboring territories to take over. So how does the visual system deal with this unfair disadvantage?

By keeping the occipital cortex active during the night.

We suggest that dreaming exists to keep the visual cortex from being taken over by neighboring areas. After all, the rotation of the planet does not affect anything about your ability to touch, hear, taste, or smell; only vision suffers in the dark. As a result, the visual cortex finds itself in danger every night of a takeover by the other senses. And given the startling rapidity with which changes in territory can happen (remember the forty to sixty minutes we just saw), the threat is formidable. Dreams are the means by which the visual cortex prevents takeover.

To better understand this, let's zoom out. Although a sleeper looks as though he is relaxed and shut down, the brain is fully electrically active. During most of the night, there is no dreaming. But during REM (rapid eye movement) sleep, something special happens. The heart rate and breathing speed up, small muscles twitch, and the brain waves become smaller and faster. This is the stage of sleep in which dreaming occurs.[41] REM sleep is triggered by a particular set of neurons in a brainstem structure called the pons. The increased activity in these neurons has two consequences. The first is that the major muscle groups become paralyzed. Elaborate neural circuitry keeps the body frozen during dreaming, and its elaborateness supports the biological importance of dream sleep; presumably, this circuitry would be unlikely to evolve without an important function behind it. The muscular shutdown allows the brain to simulate world experience without actually moving the body around.

The second consequence is the really important one: waves of

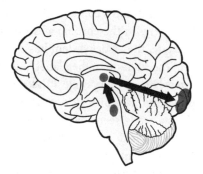

During dream sleep, waves of activity begin in the brainstem and end in the occipital cortex. We suggest this infusion of activity is necessitated by the rotation of the planet into darkness: the visual system needs special strategies to keep its territory intact.

spikes travel from the brainstem to the occipital cortex.[42] When the spikes arrive there, the activity is experienced as visual. We *see*. This activity is why dreams are pictorial and filmic, instead of conceptual or abstract.

This combination crafts the experience of dreaming: the invasion of the electrical waves into the occipital cortex makes the visual system active, while the muscular paralysis keeps the dreamer from acting on the experiences.

We theorize that the circuitry behind visual dreams is not accidental. Instead, to prevent takeover, the visual system is forced to fight for its territory by generating bursts of activity when the planet rotates into darkness.[43] In the face of constant competition for sensory real estate, an occipital self-defense evolved. After all, vision carries mission-critical information, but it is stolen away for half of our hours. Dreams, therefore, may be the strange love child of neural plasticity and the rotation of the planet.

A key point to appreciate is that these nighttime volleys of activity are anatomically precise. They begin in the brainstem and are directed to only one place: the occipital cortex. If the circuitry grew its branches broadly and promiscuously, we'd expect it to connect with many areas throughout the brain. But it doesn't. It aims with anatomical exactitude at one area alone: a tiny structure called the lateral geniculate nucleus, which broadcasts specifically to the occipital cortex. Through the neuroanatomist's lens, this high specificity of the circuit suggests an important role.

From this perspective, it should be no surprise that even a person born blind retains the same brainstem-to-occipital-lobe circuitry as everyone else. What about the dreams of blind people? Would they be expected to have no dreaming at all because their brains don't care about darkness? The answer is instructive. People who have been blind from birth (or were blinded at a very young age) experience no visual imagery in their dreams, but they do have *other* sensory experiences, such as feeling their way around a rearranged living room or hearing strange animals barking.[44] This matches perfectly with the lessons we learned a moment ago: that the occipital cortex of a blind person becomes annexed by the other senses. Thus, in the congenitally blind, nighttime occipital activation still occurs, but it is now experienced as something *nonvisual*. In other words, under normal circumstances, your genetics expect that the unfair disadvantage of darkness is best combated by sending waves of activity at night to the occipital lobe; this holds true in the brain of the blind, even though the original purpose is lost. Note also that people who become blind *after* the age of seven have more visual content in their dreams than those who become blind earlier—consistent with the fact that the occipital lobe in the late-blind is less fully conquered by other senses, and so the activity is experienced more visually.[45]

As an interesting side note, two other brain areas, the hippocampus and the prefrontal cortex, are less active during dream sleep than during the waking state, and this presumably accounts for our difficulty remembering our dreams. Why does your brain shut down these areas? One possibility is that there is no need to write memory if the central purpose of dream sleep is to keep the visual cortex actively fighting off its neighbors.

We can learn a great deal from a cross-species perspective. Some mammals are born *immature*—meaning they're unable to walk, get food, regulate their own temperature, or defend themselves. Examples are humans, ferrets, and platypuses. Other mammals are born *mature*—such as the guinea pig, sheep, and giraffe—all of whom come out of the womb with teeth, fur, open eyes, and an ability to regulate their temperature, walk within an hour of being born, and eat solid food. Here's the important clue: the animals born immature have much more REM sleep—up to about eight times as much—and this difference is especially clear in the first month of life.[46] In our interpretation, when a highly plastic brain drops into the world, it needs to constantly fight to keep things balanced. When a brain arrives mostly solidified, there is less need for it to engage in the nighttime fighting.

Moreover, look at the falloff in REM sleep with age. All mammalian species spend some fraction of their sleep time in REM, and that fraction steadily decreases as they get older.[47] In humans, infants spend half of their sleeping time in REM, adults spend only 10–20 percent of sleep in REM, and the elderly spend even less. This cross-species trend is consistent with the fact that infants' brains are so much more plastic (as we will see in chapter 9), and thus the competition for territory is even more critical. As an animal gets older, cortical takeovers become less possible. The falloff in plasticity parallels the falloff of time spent in REM sleep.

This hypothesis leads to a prediction for the distant future, when we discover life on other planets. Some planets (especially those orbiting red dwarf stars) become locked into place, such that they always have the same surface facing their star: they thus have permanent day on one side of the planet, and permanent night on the other.[48] If life-forms on that planet were to have livewired brains even vaguely similar to ours, the prediction would be that those on the daylight side of the planet might have vision like us but would *not* have dreams. The same prediction would apply for very fast-spinning planets: if the nighttime is shorter than the time of a cortical takeover, then dreaming would also be unnecessary. Thousands of years hence, we might finally know whether we dreamers are in the universal minority.

AS OUTSIDE, SO INSIDE

Most visitors to Admiral Nelson's statue in London's Trafalgar Square have probably not considered the distortion of the somatosensory cortex in the left hemisphere of that elevated head. But they should. It exposes one of the most remarkable feats of the brain: the ability to optimally encode the body it is dealing with.

We've seen so far that changes to sensory inputs (as with amputation or blindness or deafness) lead to massive cortical reorganization. The brain's maps are not genetically pre-scripted but instead molded by the input. They are experience-dependent. They are an emergent property of local border competitions rather than the result of a pre-specified global plan. Because neurons that fire together wire together, co-activation establishes neighboring representations in the brain. No matter the shape of your body, it will naturally end up mapped on the brain's surface.

Evolutionarily, such activity-dependent mechanisms allow natural selection to quickly test out innumerable varieties of body types—from claws to fins, wings to prehensile tails. Nature does not need to genetically rewrite the brain each time it wants to try out a new body plan; it simply lets the brain adjust itself. And this underscores a point that reverberates throughout this book: the brain is very different from a digital computer. We'll want to abandon our notions of traditional engineering and keep our eyes wide open as we move deeper into the neural terrain.

The shape shifting around the body plan illustrates what happens in all sensory systems. We saw that when people are born blind, their "visual" cortex becomes tuned to hearing, touch, and other senses. And the perceptual consequence of the cortical takeover is increased sensitivity: the more real estate the brain devotes to a task, the higher resolution it has.

Finally, we discovered that when people with normal visual systems are blindfolded for as little as an hour, their primary visual cortex becomes active when they perform tasks with their fingers or when they hear tones or words. Removing the blindfold quickly reverts the visual cortex so that it responds only to visual input. As we'll discover in upcoming chapters, the brain's sudden ability to "see" with the fingers and ears depends on connections from other senses that are already there but not used so long as the eyes are sending data.

Collectively, these considerations led us to propose that visual dreams are a by-product of neural competition and the rotation of the planet. An organism that wishes to keep its visual system from takeover by the other senses must devise a way to keep the visual system active when the darkness sets in.

So now we're ready for a question. We've painted the picture of an extremely flexible cortex. What are the limits of its flexibility? Can we feed any kinds of data into the brain? Would it simply figure out what to do with the data it receives?

4

WRAPPING AROUND THE INPUTS

Every man can, if he so desires, become the sculptor of his own brain.

—SANTIAGO RAMÓN Y CAJAL (1852–1934),
neuroscientist and Nobel laureate

Michael Chorost was born with poor hearing, and he got by during his young adult life with the help of a hearing aid. But one afternoon, while waiting to pick up a rental car, the battery to his hearing aid died. Or so he thought. He replaced the battery but found that all sound was still missing from his world. He drove himself to the nearest emergency room and discovered that the remainder of his hearing—his thin auditory lifeline to the rest of the world—was gone for good.[1]

Hearing aids wouldn't be of any use for him now; after all, they work by capturing sound from the world and blasting it at higher volume into the ailing auditory system. This strategy is effective for some types of hearing loss, but it only works if everything downstream of the eardrum is functioning. If the inner ear is defunct, no amount of amplification solves the problem. And this was Michael's situation. It seemed as though his perception of the world's soundscapes had come to an end.

But then he found out about a single remaining possibility, and in 2001 he underwent surgery for a cochlear implant. This tiny device circumvents the broken hardware of the inner ear to speak directly to the functioning nerve (think of it like a data cable) just beyond it. The implant is a minicomputer lodged directly into his inner ear; it receives sounds from the outside world and passes the information to the auditory nerve by means of tiny electrodes.

So the damaged part of the inner ear is bypassed, but that doesn't mean the experience of hearing comes for free. Michael had to learn to interpret the foreign language of the electrical signals being fed to his auditory system:

> When the device was turned on a month after surgery, the first sentence I heard sounded like "Zzzzzz szz szvizzz ur brfzzzzzz?" My brain gradually learned how to interpret the alien signal. Before long, "Zzzzzz szz szvizzz ur brfzzzzzz?" became "What did you have for breakfast?" After months of practice, I could use the telephone again, even converse in loud bars and cafeterias.

Although being implanted with a minicomputer sounds something like science fiction, cochlear implants have been on the market since 1982, and more than half a million people are walking around with these bionics in their heads, enjoying voices and door knocks and laughter and piccolos. The software on the cochlear implant is hackable and updateable, so Michael has spent years getting more efficient information through the implant without further surgeries. Almost a year after the implant was activated, he upgraded to a program that gave him twice the resolution. As Michael puts it, "While my friends' ears will inevitably decline with age, mine will only get better."

Terry Byland lives near Los Angeles. He was diagnosed with retinitis pigmentosa, a degenerative disorder of his retina, the sheet of photoreceptors at the back of the eye. He reports, "Aged 37, the last thing you want to hear is that you are going blind—that there's nothing they can do."[2]

But then he discovered that there *was* something he could do, if he was brave enough to try it. In 2004, he became one of the first patients to undergo an experimental procedure: getting implanted with a bionic retinal chip. A tiny device with a grid of electrodes, it plugs into the retina at the back of the eye. A camera on glasses wirelessly beams its signals to the chip. The electrodes give little zaps of electricity to Terry's surviving retinal cells, generating signals along the previously silent highway of the optic nerve. After all, Terry's optic nerve functioned just fine: even while the photoreceptors had died, the nerve remained hungry for signals it could carry to the brain.

A research team at the University of Southern California implanted the miniature chip in Terry's eye. The surgery was completed without a hitch, and then the real testing began. With hushed anticipation, the research team turned on the electrodes individually to test them. Terry reported, "It was amazing to see something. It was like little specks of light—not even the size of a dime—when they were testing the electrodes one by one."

Over the course of days, Terry experienced only small constellations of lights: not a rousing success. But his visual cortex gradually figured out how to extract better information out of the signals. After some time, he detected the presence of his eighteen-year-old son: "I was with my son, walking . . . it was the first time I had seen him since he was five years old. I don't mind saying, there were a few tears wept that day."

Terry wasn't experiencing a clear visual picture—it was more like a simple pixelated grid—but the door of darkness had swung open a crack. Over time, his brain has been able to make better sense of the signals. While he can't ascertain the details of individual faces, he can make them out dimly. And although the resolution of his retinal chip is low, he can touch objects presented at random locations and is able to cross a city street by discerning the white lines of the crosswalk.[3] He proudly reports, "When I'm in my home, or another person's house, I can go into any room and switch the light on, or see the light coming in through the window. When I am walking along the street I can avoid low hanging branches—I can see the edges of the branches, so I can avoid them."

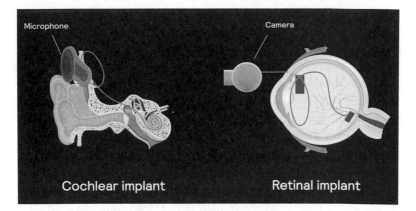

These digital devices push information that doesn't quite match the language of the natural biology. Nevertheless, the brain figures out how to make use of the data.

The idea of prostheses for the ear and eye had been seriously considered in the scientific community for decades. But no one was positive that these technologies would work. After all, the inner ear and the retina perform astoundingly sophisticated processing on the sensory input they receive. So would a small electronic chip, speaking the dialect of Silicon Valley instead of the language of our natural biological sense organs, be understood by the rest of the brain? Or instead, would its patterns of miniature electrical sparks come off as gibberish to downstream neural networks? These devices would be like an uncouth traveler to a foreign land who expects that everyone will figure out his language if he just keeps shouting it.

Amazingly, in the case of the brain, such an unrefined strategy works: the rest of the country learns to understand the foreigner.

But how?

The key to understanding this requires diving one level deeper: your three pounds of brain tissue are not directly hearing or seeing any of the world around you. Instead, your brain is locked in a crypt of silence and darkness inside your skull. All it ever sees are electrochemical signals that stream in along different data cables. That's all it has to work with.

In ways we are still working to understand, the brain is stunningly gifted at taking in these signals and extracting patterns. To those patterns it assigns meaning. With the meaning you have subjective experience. The brain is an organ that converts sparks in the dark into the euphonious picture show of your world. All of the hues and aromas and emotions and sensations in your life are encoded in trillions of signals zipping in blackness, just as a beautiful screen saver on your computer screen is fundamentally built of zeros and ones.

THE PLANET-WINNING TECHNOLOGY OF THE POTATO HEAD

Imagine you went to an island of people born blind. They all read by Braille, feeling tiny patterns of inputs on their fingertips. You watch them break into laughter or melt into sobs as they brush over the small bumps. How can you fit all that emotion into the tip of your finger? You explain to them that when *you* enjoy a novel, you aim the spheres on your face toward particular lines and curves. Each sphere has a lawn of cells that record collisions with photons, and in this way you can register the shapes of the symbols. You've memorized a set of rules by which different shapes represent different sounds. Thus, for each squiggle you recite a small sound in your head, imagining what you would hear if someone were speaking aloud. The resulting pattern of neurochemical signaling makes you explode with hilarity or burst into tears. You couldn't blame the islanders for finding your claim difficult to understand.

You and they would finally have to allow a simple truth: the fingertip or the eyeball is just the peripheral device that converts information from the outside world into spikes in the brain. The brain then does all the hard work of interpretation. You and the islanders would break bread over the fact that in the end it's all about the trillions of spikes racing around in the brain—and that the method of entry simply isn't the part that matters.

Whatever information the brain is fed, it will learn to adjust to it and extract what it can. As long as the data have a structure that reflects something important about the outside world (along with some other

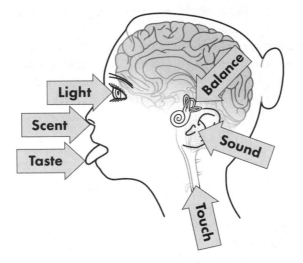

Sensory organs feed many different information sources to the brain.

requirements we will see in the next chapters), the brain will figure out how to decode it.

There's an interesting consequence to this: your brain doesn't know, and it doesn't care, where the data come from. Whatever information comes in, it just works out how to leverage it.

This makes the brain a very efficient kind of machine. It is a general-purpose computing device. It just sucks up the available signals and determines—nearly optimally—what it can do with them. And that strategy, I propose, frees up Mother Nature to tinker around with different sorts of input channels.

I call this the Potato Head model of evolution. I use this name to emphasize that all the sensors that we know and love—like our eyes and our ears and our fingertips—are merely peripheral plug-and-play devices. You stick them in, and you're good to go. The brain figures out what to do with the data that come in.

As a result, Mother Nature can build new senses simply by building new peripherals. In other words, once she has figured out the operating principles of the brain, she can tinker around with different sorts of input channels to pick up on different energy sources from

*The Potato Head hypothesis: plug in sensory organs,
and the brain figures out how to use them.*

the world. Information carried by the reflection of electromagnetic radiation is captured by the photon detectors in the eyes. Air compression waves are captured by the sound detectors of the ears. Heat and texture information is gathered by the large sheets of sensory material we call skin. Chemical signatures are sniffed or licked up by the nose or tongue. And it all gets translated into spikes running around in the dark vault of the skull.

This remarkable ability of the brain to accept any sensory input shifts the burden of research and development of new senses to the exterior sensors. In the same way that you can plug in an arbitrary nose or eyes or mouth for Potato Head, likewise does nature plug a wide variety of instruments into the brain for the purpose of detecting energy sources in the outside world.

Consider plug-and-play peripheral devices for your computer. The importance of the designation "plug-and-play" is that your computer does not have to know about the existence of the XJ-3000 SuperWeb-Cam that will be invented several years from now; instead, it needs only to be open to interfacing with an unknown, arbitrary device and receiving streams of data when the new device gets plugged in. As a result, you do not need to buy a new computer each time a new peripheral hits the market. You simply have a single, central device that opens its portholes for peripherals to be added in a standardized manner.[4]

Viewing our peripheral detectors like individual, stand-alone devices might seem crazy; after all, aren't thousands of genes involved in building these devices, and don't those genes overlap with other pieces and parts of the body? Can we really look at the nose, eye, ear, or tongue as a device that stands alone? I dove deep into researching this problem. After all, if the Potato Head model was correct, wouldn't that suggest that we might find simple switches in the genetics that lead to the presence or absence of these peripherals?

As it turns out, all genes aren't equal. Genes unpack in an exquisitely precise order, with the expression of one triggering the expression of the next ones in a sophisticated algorithm of feedback and feedforward. As a result, there are critical nodes in the genetic program for building, say, a nose. That program can be turned on or off.

How do we know this? Look at mutations that happen with a genetic hiccup. Take the condition called arhinia, in which a child is born without a nose. It is simply missing from the face. Baby Eli, born in Alabama in 2015, is completely missing a nose, and he also lacks a nasal cavity or system for smelling.[5] Such a mutation seems startling and difficult to fathom, but in our plug-and-play framework arhinia is predictable: with a slight tweak of the genes, the peripheral device simply doesn't get built.

Baby Eli was born with no nose.

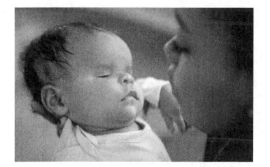

Baby Jordy was born with no eyes; beneath his lids one finds skin.

If our sensory organs can be viewed as plug-and-play devices, we might expect to find medical cases in which a child is born with, say, no eyes. And indeed, that is exactly what the condition of anophthalmia is. Consider baby Jordy, born in Chicago in 2014.[6] Beneath his eyelids, one simply finds smooth, glossy flesh. While Jordy's behavior and brain imaging indicate that the rest of his brain is functioning just fine, he has no peripheral devices for capturing photons. Jordy's grandmother points out, "He will know us by feeling us." His mother, Brania Jackson, got a special "I love Jordy" tattoo—in Braille—on her right shoulder blade so that Jordy can grow up feeling it.

Some babies are born without ears. In the rare condition of anotia, children are born with a complete absence of the external portion of the ear.

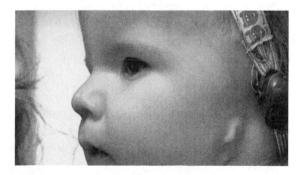

A child with no ears.

Relatedly, a mutation of a single protein makes the structures of the *inner* ear absent.[7] Needless to say, children with these mutations are completely deaf, because they lack the peripheral devices that convert air pressure waves into spikes.

Can you be born without a tongue, but otherwise healthy? Sure. That's what happened to a Brazilian baby named Auristela. She spent years struggling to eat, speak, and breathe. Now an adult, she underwent an operation to put in a tongue, and at present she gives eloquent interviews on growing up tongueless.[8]

The extraordinary list of the ways we can be disassembled goes on. Some children are born without any pain receptors in their skin and inner organs, so they are totally insensitive to the sting and agony of life's lesser moments.[9] (At first blush, it might seem as though freedom from pain would be an advantage. But it's not: children unable to experience pain are covered with scars and often die young because they don't know what to avoid.) Beyond pain, there are many other types of receptors in the skin, including stretch, itch, and temperature, and a child can end up missing some, but not others. This collectively falls under the term "anaphia," the inability to feel touch.

When we look at this constellation of disorders, it becomes clear that our peripheral detectors unpack by dint of specific genetic programs. A minor malfunction in the genes can halt the program, and then the brain doesn't receive that particular data stream.

———————

The all-purpose cortex idea suggests how new sensory skills can be added during evolution: with a mutated peripheral device, a new data stream makes its way into some swath of the brain, and the neural processing machinery gets to work. Thus, new skills require only the development of new sensory devices.

And that's why we can look across the animal kingdom and find all kinds of strange peripheral devices, each of which is crafted by millions of years of evolution. If you were a snake, your sequence of DNA would fabricate heat pits that pick up infrared information. If you were a black ghost knifefish, your genetic letters would unpack electrosensors that pick up on perturbations in the electrical field. If you were a bloodhound dog, your code would write instructions for an enormous snout crammed with smell receptors. If you were a mantis shrimp, your instructions would manufacture eyes with sixteen types of photoreceptors. The star-nosed mole has twenty-two finger-like appendages on its nose, and with these it feels around and constructs a 3-D model of its tunnel systems. Many birds, cows, and insects have magnetoreception, with which they orient to the magnetic field of the planet.

To accommodate such varied peripherals, does the brain have to be redesigned each time? I suggest not. In evolutionary time, random mutations introduce strange new sensors, and the recipient brains simply figure out how to exploit them. Once the principles of brain operation have been established, nature can simply worry about designing new sensors.

This perspective allows a lesson to come into focus: the devices we come to the table with—eyes, noses, ears, tongues, fingertips—are not the only collection of instruments we could have had. These are simply what we've inherited from a lengthy and complex road of evolution.

But that particular collection of sensors may not be what we have to stick with.

After all, the brain's ability to wrap itself around different kinds of incoming information implies the bizarre prediction that you might be able to get one sensory channel to carry another's information. For example, what if you took a data stream from a video camera and converted it into touch on your skin? Would the brain eventually be able to interpret the visual world simply by feeling it?

Welcome to the stranger-than-fiction world of sensory substitution.

SENSORY SUBSTITUTION

The idea that one can feed data into the brain via the wrong channels may sound hypothetical and bizarre. But the first paper demonstrating this was published in the journal *Nature* more than half a century ago.

The story begins in 1958, when a physician named Paul Bach-y-

Sensory substitution: feed information into the brain via unusual pathways.

Rita received terrible news: his father, a sixty-five-year-old professor, had just suffered a major stroke. He was wheelchair bound and could barely speak or move. Paul and his brother George, a medical student at the University of Mexico, searched for ways to help their father. And together they pioneered an idiosyncratic, one-on-one rehabilitation program.

As Paul described it, "It was tough love. [George would] throw something on the floor and say 'Dad, go get it.' "[10] Or they would have their father try to sweep the porch, even as the neighbors looked on in dismay. But for their father, the struggle was rewarding. As Paul phrased his father's view, "This useless man was doing something."

Stroke victims frequently recover only partially—and often not at all—so the brothers tried not to buy into false hope. They knew that when brain tissue is killed in a stroke, it never comes back.

But their father's recovery proceeded unexpectedly well. So well, in fact, that their father returned to his professorship and died much later in life (the victim of a heart attack while hiking in Colombia at nine thousand feet).

Paul was deeply impressed at the extent of his father's recovery, and the experience marked a major turning point in his life. Paul realized that the brain could retrain itself and that even when parts of the brain were forever gone, other parts could take over their function. Paul departed a professorship at Smith-Kettlewell in San Francisco to begin a residency in rehabilitation medicine at Santa Clara Valley Medical Center. He wanted to study people like his father. He wanted to figure out what it took for the brain to retrain.

By the late 1960s, Paul Bach-y-Rita had pursued a scheme that most of his colleagues assumed to be foolish. He sat a blind volunteer in a reconfigured dental chair in his laboratory. Inset into the back of the chair was a grid of four hundred Teflon tips in a twenty-by-twenty grid. The tips could be extended and retracted by mechanical solenoids. Over the blind man's head a camera was mounted on a tripod. The video stream of the camera was converted into a poking of the tips against the volunteer's back.

Objects were passed in front of the camera while the blind participant in the chair paid careful attention to the feelings in his back. Over days of training, he got better at identifying the objects by their feel—in the same way a person might play the game of drawing with a finger on another person's back and then asking the person to identify the shape or letter. The experience wasn't exactly like *vision,* but it was a start.

What Bach-y-Rita found astonished the field: the blind subjects

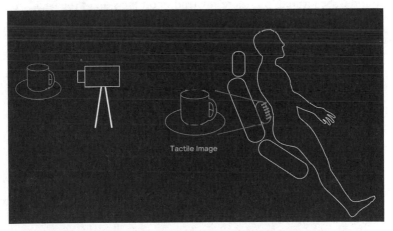

Tactile Image

A video feed is translated into touch on the back.

could learn to distinguish horizontal from vertical from diagonal lines. More advanced users could learn to distinguish simple objects and even faces—simply by the poking sensations on their back. He published his findings in the journal *Nature,* under the surprising title "Vision Substitution by Tactile Image Projection." It was the beginning of a new era—that of sensory substitution.[11] Bach-y-Rita summarized his findings simply: "The brain is able to use information coming from the skin as if it were coming from the eyes."

The technique improved drastically when Bach-y-Rita and his collaborators made a single, simple change: instead of mounting the camera to the chair, they allowed the blind user to point it himself, using his own volition to control where the "eye" looked.[12] Why? Because sensory input is best learned when one can interact with the world. Letting users control the camera closed the loop between muscle output and sensory input.[13] Perception can be understood not as passive but instead as a way to actively explore the environment, matching a particular action to a specific change in what returns to the brain. It doesn't matter to the brain how that loop gets established—whether by moving the extraocular muscles that move the eye or using arm muscles to tilt a camera. However it happens, the brain works to figure out how the output maps to the input.

The subjective experience for the users was that visual objects were found to be located "out there" instead of on the skin of the back.[14] In other words, it was something like vision. Even though the sight of your friend's face at the coffee shop impinges on your photoreceptors, you don't perceive that the signal is at your eyes. You perceive that he's *out there,* waving at you from a distance. And so it went for the users of the modified dental chair.

While Bach-y-Rita's device was the first to hit the public eye, it was not actually the first attempt at sensory substitution. On the other side of the world at the end of the 1890s, a Polish ophthalmologist named Kazimierz Noiszewski developed the Elektroftalm (from the Greek for "electricity" + "eye") for blind people. A photocell was placed on the forehead of a blind person, and the more light that hit it, the louder a sound in the person's ear. Based on the sound's intensity, the blind person could tell where lights or dark areas were.

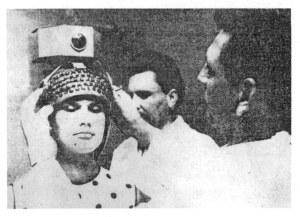

The Elektroftalm translated a camera's image into vibrations on the head (1969).

Unfortunately, it was large, heavy, and only one pixel of resolution, so it gained no traction. But by 1960, his Polish colleagues picked up the ball and ran with it.[15] Recognizing that hearing is critical for the blind, they instead turned to passing in the information via touch. They built a system of vibratory motors, mounted on a helmet, that "drew" the images on the head. Blind participants could move around in specially prepared rooms, painted to enhance the contrast of door frames and furniture edges. It worked. Alas, like the earlier inventions, the device was heavy and would get hot during use, so the world had to wait. But the proof of principle was there.

Why did these strange approaches work? Because inputs to the brain—photons at the eyes, air compression waves at the ears, pressure on the skin—are all converted into the common currency of electrical signals. As long as the incoming spikes carry information that represents something important about the outside world, the brain will learn how to interpret it. The vast neural forests in the brain don't care about the route by which the spikes entered. Bach-y-Rita described it this way in a 2003 interview on PBS:

> If I'm looking at you, the image of you doesn't get beyond my retina. . . . From there to the brain to the rest of the brain, it's pulses. Pulses along nerves. Those pulses aren't any different from the

pulses along the big toe. It's [the] information that [they carry], and the frequency and the pattern of pulses. If you could train the brain to extract that kind of information, then you don't need the eye to see.

In other words, the skin is a path to feeding data into a brain that no longer possesses functioning eyes. But how could that work?

THE ONE-TRICK PONY

When you look at the cortex, it looks approximately the same everywhere as you traverse its hills and valleys. But when we image the brain or dip tiny electrodes into its jellylike mass, we find that different types of information are lurking in different regions. These differences have allowed neuroscientists to assign areas with labels: this region is for vision, this one for hearing, this one for touch from your left big toe, and so on. But what if areas come to be what they are only because of their inputs? What if the "visual" cortex is only visual because of the data it receives? What if specialization develops from the details of the incoming data cables rather than by genetic pre-specification of modules? In this framework, the cortex is an all-purpose data-processing engine. Feed data in and it will crunch through and extract statistical regularities.[16] In other words, it's willing to accept whatever input is plugged into it and performs the same basic algorithms on it. In this view, no part of the cortex is prespecified to be visual, auditory, and so on. So whether an organism wants to detect air compression waves or photons, all it has to do is plug the fiber bundle of incoming signals into the cortex, and the six-layered machinery will run a very general algorithm to extract the right kind of information. The data make the area.

And this is why the neocortex looks about the same everywhere: because it *is* the same. Any patch of cortex is pluripotent—meaning that it has the possibility to develop into a variety of fates, depending on what's plugged into it.

So if there's an area of the brain devoted to hearing, it's only because peripheral devices (in this case, the ears) send information along cables that plug into the cortex at that spot. It's not the auditory cortex by necessity; it's the auditory cortex only because signals passed along by the ears have shaped its destiny. In an alternate universe, imagine that nerve fibers carrying visual information plugged into that area; then we would label it in our textbooks as the visual cortex. In other words, the cortex performs standard operations on whatever input it happens to get. This gives a first impression that the brain has prespecified sensory areas, but it really only looks that way because of the inputs.[17]

Consider where the fish markets are in the middle United States: the towns in which pescatarianism thrives, in which sushi restaurants are overrepresented, in which new seafood recipes are developed— let's call these towns the primary fishual areas.

Why does the map have a particular configuration, and not something different? It looks that way because that's where the rivers flow, and therefore where the fish are. Think of the fish like bits of data, flowing along the data cables of the rivers, and the restaurant distribution crafts itself accordingly. No legislative body prescribed that the fish markets should move there. They clustered there naturally.

All this leads to the hypothesis that there's nothing special about a chunk of tissue in, say, the auditory cortex. So could you cut out a bit of auditory cortex in an embryo and transplant it into the visual cortex, and would it function just fine? Indeed, this is precisely what was demonstrated in animal experiments beginning in the early 1990s: in short order, the chunk of transplanted tissue looks and behaves just like the rest of the visual cortex.[18]

And then the demonstration was taken a step further. In 2000, scientists at MIT redirected inputs from a ferret's eye to the auditory cortex so that now the *auditory* cortex received *visual* data. What happened? The auditory cortex adjusted its circuitry to resemble the connections of the primary visual cortex.[19] The rewired animals interpreted inputs to the auditory cortex like normal vision. This tells us that the pattern of inputs determines the fate of the cortex. The brain dynamically wires itself to best represent (and eventually act upon) whatever data come swimming in.[20]

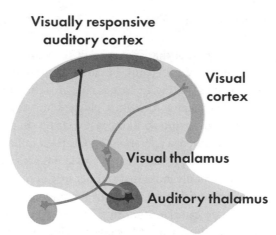

Visual fibers in the ferret brain were rerouted to the auditory cortex—which then began to process visual information.

Hundreds of studies on transplanting tissue or rewiring inputs support the model that the brain is a general-purpose computing device—a machine that performs standard operations on the data streaming in—whether those data carry a glimpse of a hopping rabbit, the sound of a phone ring, the taste of peanut butter, the smell of salami, or the touch of silk on the cheek. The brain analyzes the input and puts it into context (*what can I do with this?*), regardless of where it comes from. And that's why data can become useful to a blind person even when they're fed into the back, or ear, or forehead.

In the 1990s, Bach-y-Rita and his colleagues sought ways to go smaller than the dental chair. They developed a small device called the Brain-Port.[21] A camera is attached to the forehead of a blind person, and a small grid of electrodes is placed on the tongue. The "Tongue Display Unit" uses a grid of stimulators over three square centimeters. The electrodes deliver small shocks that correlate with the position of pixels, feeling something like the children's candy Pop Rocks in the mouth. Bright pixels are encoded by strong stimulation at the corresponding points on the tongue, gray by medium stimulation, and

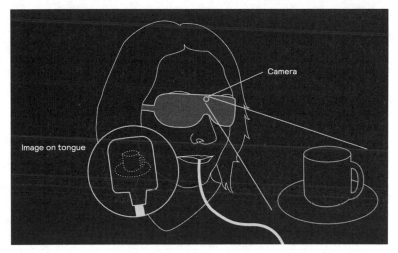

Seeing with the tongue.

darkness by no stimulation. The BrainPort gives the capacity to distinguish visual items with a visual acuity that equates to about 20/800 vision.[22] While users report that they first perceived the tongue stimulation as unidentifiable edges and shapes, they eventually learn to recognize the stimulation at a deeper level, allowing them to discern qualities such as distance, shape, direction of movement, and size.[23]

We normally think of the tongue as a taste organ, but it is loaded with touch receptors (that's how you feel the texture of food), making it an excellent brain-machine interface.[24] As with the other visual-tactile devices, the tongue grid reminds us vision arises not in the eyes but in the brain. When brain imaging is performed on trained subjects (blind or sighted), the motion of electrotactile shocks across the tongue activates an area of the brain normally involved in visual motion.[25]

As with the grid of solenoids on the back, blind people who use the BrainPort begin to feel that scenes have "openness" and "depth" and that objects are *out there*. In other words, it's more than a cognitive translation of what's happening on the tongue: it grows into a direct perceptual experience. Their experience is not "I feel a pattern on my tongue that codes for my spouse passing by," but instead a direct sense that their spouse is moving across the living room. If you have normal vision, keep in mind this is precisely how your eyes work: electrochemical signals in your retinas are perceived as a friend beckoning you, a Ferrari zooming past on the road, a scarlet kite against an azure sky. Even though all the activity is at the surface of your sensory detectors, you perceive everything as *out there*. It simply doesn't matter whether the detector is the eye or the tongue. As the blind participant Roger Behm describes his experience of the BrainPort:

> Last year, when I was up here for the first time, we were doing stuff on the table, in the kitchen. And I got kind of a little emotional, because it's thirty-three years since I've seen before. And I could reach out and I see the different-sized balls. I mean I visually see them. I could reach out and grab them—not grope or feel for them—pick them up, and see the cup, and raise my hand and drop it right in the cup.[26]

As you can presumably guess by now, the tactile input can be almost anywhere on the body. Researchers in Japan have developed a variant of the tactile grid—the Forehead Retina System—in which a video

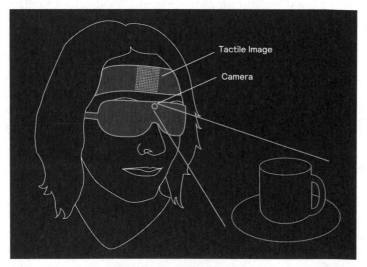

The Forehead Retina System.

stream is converted to small points of touch on the forehead.[27] Why the forehead? Why not? It's not being used for much else.

Another version hosts a grid of vibrotactile actuators on the abdomen, which use intensity to represent distance to the nearest surfaces.[28]

What these all have in common is that the brain can figure out what to make of visual input coming in through channels normally thought of as touch. But it turns out that touch isn't the only strategy that works.

EYE TUNES

In my laboratory some years ago, Don Vaughn walked with his iPhone held out in front of him. His eyes were closed, and yet he was not crashing into things. The sounds streaming through his earbuds were busily converting the visual world into a soundscape. He was learning to see the room with his ears. He gently moved the phone around in front of him like a third eye, like a miniature walking cane, turning it this way and that to pull in the information he needed. We were testing whether a blind person could pick up visual information through the ears. Although you might not have heard of this approach to blindness before, the idea isn't new: it began more than half a century earlier.

In 1966, a professor named Leslie Kay became obsessed with the beauty of bat echolocation. He knew that some humans could learn to echolocate, but it wasn't easy. So Kay designed a bulky pair of glasses to help the blind community take advantage of the idea.[29]

The glasses emitted an ultrasonic sound into the environment. With its short wavelengths, ultrasonic can reveal information about small objects when it bounces back. Electronics on the glasses captured the returning reflections and converted them into sounds humans could hear. The note in your ear indicated the distance of the object: high pitches coded for something far away, low pitches for something nearby. The volume of a signal told you about the size of the object: loud meant the object was large; soft told you it was small. The clarity of the signal was used to represent texture: a smooth object became a

Professor Kay's sonic glasses shown on the right.
(The other glasses are merely thick, not sonic.)

pure tone; a rough texture sounded like a note corrupted with noise. Users learned to perform object avoidance pretty well; however, because of the low resolution, Kay and his colleagues concluded that the invention served more as a supplement to a guide dog or cane rather than a replacement.

Although it was only moderately useful for adults, there remained the question of how well a baby's brain might learn to interpret the signals, given that young brains are especially plastic. In 1974, in California, the psychologist T. G. R. Bower used a modified version of Kay's glasses to test whether the idea could work. His participant was a sixteen-week-old baby, blind from birth.[30] On the first day, Bower took an object and moved it slowly back and forth from the infant's nose. By the fourth time he moved the object, he reports, the baby's eyes converged (both pointed toward the nose), as happens when something approaches the face. When Bower moved the object away, the baby's eyes diverged. After a few more cycles of this, the baby put up its hands as the object drew near. When objects were moved left and right in front of the baby, Bower reports that the baby tracked them with its head and tried to swipe at them. In his write-up of the results, Bower relates several other behaviors:

> The baby was facing [his talking mother] and wearing the device. He slowly turned his head to remove her from the sound field, then slowly turned back to bring her in again. This behavior was repeated several times to the accompaniment of immense smiles from the baby. All three observers had the impression that he was playing a kind of peek-a-boo with his mother, and deriving immense pleasure from it.

He goes on to report remarkable results over the next several months:

> The baby's development after these initial adventures remained more or less on a par with that of a sighted baby. Using the sonic guide the baby seemed able to identify a favorite toy without touching it. He began two-handed reaches around 6 months of age. By 8 months the baby would search for an object that had been hidden behind another object. . . . None of these behavior patterns is normally seen in congenitally blind babies.

You may wonder why you haven't heard of these being used before. Just as we saw earlier, the technology was bulky and heavy—not the kind of thing you could reasonably grow up using—while the resolution was fairly low. Further, the results in adults generally met with less success than those in children with the ultrasonic glasses[31]—an issue we'll return to in chapter 9. So while the concept of sensory substitution took root, it had to wait for the right combination of factors to thrive.

―――――――――

In the early 1980s, a Dutch physicist named Peter Meijer picked up the baton of thinking about the ears as a means to transmit visual information. Instead of using echolocation, he wondered if he could take a video feed and convert it into sound.

He had seen Bach-y-Rita's conversion of a video feed into touch, but he suspected that the ears might have a greater capacity to soak in information. The downside of going for the ears was that the conversion from video to sound was going to be less intuitive. In Bach-y-Rita's dental chair, the shape of a circle, face, or person could be pressed directly against the skin. But how does one convert hundreds of pixels of video into sound?

By 1991, Meijer had developed a version on a desktop computer, and by 1999 it was portable, worn as camera-mounted glasses with a computer clipped to the belt. He called his system the vOICe (where "OIC" stands for "Oh, I See").[32] The algorithm manipulates sound along three dimensions: the height of an object is represented by the *frequency* of the sound, the horizontal position is represented by time via a panning of the stereo input (imagine sound moving in the ears from left to right, the way you scan a scene with your eyes), and the brightness of an object is represented by volume. Visual information could be captured for a grayscale image of about sixty by sixty pixels.[33]

Try to imagine the experience of using these glasses. At first, everything sounds like a cacophony. As one moves around the environment, pitches are buzzing and whining in an alien and useless manner. After a while, one gets a sense of how to use the sounds to navigate around. At this stage it is a cognitive exercise: one is laboriously translating the pitches into something that can be acted upon.

The important part comes a little later. After some weeks or months, blind users begin performing well.[34] But not just because they have memorized the translation. Instead, they are, in some sense, *seeing*. In a strange, low-resolution way, they are experiencing vision.[35] One of the vOICe users, who went blind after twenty years of vision, had this to say about her experience of using the device:

> You can develop a sense of the soundscapes within two to three weeks. Within three months or so you should start seeing the flashes of your environment where you'd be able to identify things just by looking at them. . . . It is sight. I know what sight looks like. I remember it.[36]

Rigorous training is key. Just as with cochlear implants, it can take many months of using these technologies before the brain starts to make sense of the signals. By that point, the changes are measurable in brain imaging. A particular region of the brain (the lateral occipital cortex) normally responds to shape information, whether the shape is determined by sight or by touch. After users wear the glasses for several days, this brain region becomes activated by the soundscape.[37] The improvements in a user's performance are paralleled by the amount of cerebral reorganization.[38]

In other words, the brain figures out how to extract shape information from incoming signals, regardless of the path by which those signals get into the inner sanctum of the skull—whether by sight, touch, or sound. The details of the detectors don't matter. All that matters is the information they carry.

By the early years of the twenty-first century, several laboratories began to take advantage of cell phones, developing apps to convert camera input into audio output. Blind people listen through their earbuds as they view the scene in front of them with the cell phone camera. For example, the vOICe can now be downloaded for free on phones around the world.

The vOICe is not the only visual-to-auditory substitution approach; recent years have seen a proliferation of these technologies. For example, the EyeMusic app uses musical pitches to represent the up-down location of pixels: the higher a pixel, the higher the note. Timing is exploited to represent the left-right pixel location: earlier notes are used for something on the left; later notes represent something on the right. Color is conveyed by different musical instruments: white (vocals), blue (trumpet), red (organ), green (reed), yellow (violin).[39] Other groups are experimenting with alternative versions: for example, using magnification at the center of the scene, just like the human eye, or using simulated echolocation or distance-dependent volume modulation, or many other ideas.[40]

The ubiquity of smartphones has moved the world from bulky computers to colossal power in the back pocket. And this allows not only efficiency and speed but also a chance for sensory-substitution devices to gain global leverage, especially as 87 percent of visually impaired people live in developing countries.[41] Inexpensive sensory-substitution apps can have a worldwide reach, because they involve no ongoing cost of production, physical dissemination, stock replenishment, or adverse medical reactions. In this way, a neurally inspired approach can be inexpensive and rapidly deployable and tackle global health challenges.

If it seems surprising that a blind person can come to "see" with her tongue or through cell phone earbuds, just remember how the blind come to read Braille. At first the experience involves mysterious bumps on the fingertips. But soon it becomes more than that: the brain moves beyond the details of the medium (the bumps) for a direct experience of the meaning. The Braille reader's experience parallels yours as your eyes flow over this text: although these letters are arbitrary shapes, you surpass the details of the medium (the letters) for a direct experience of the meaning.

To the first-time wearer of the tongue grid or the sonic headphones, the data streaming in require translation: the signals generated by a visual scene (say, a dog entering the living room with a bone in her mouth) give little indication of what's out there. It's as though the nerves were passing messages in a foreign language. But with enough practice, the brain can learn to translate. And once it does, the understanding of the visual world becomes directly apparent.

GOOD VIBRATIONS

Given that 5 percent of the world has disabling hearing loss, researchers some years ago got interested in ferreting out the genetics involved.[42] Unfortunately, the community has currently discovered more than 220 genes associated with deafness. For those hoping for a simple solution, this is a disappointment, but it's no surprise. After all, the auditory system works as a symphony of many delicate pieces operating in concert. As with any complex system, there are hundreds of ways it can be disrupted. As soon as any part of the system goes wrong, the whole system suffers, and the result is lumped into the term "hearing loss."

Many researchers are working to figure out how to repair those individual pieces and parts. But let's ask the question from the livewiring point of view: How could the principles of sensory substitution help us solve the problem?

With this question in mind, my former graduate student Scott Novich and I wanted to build sensory substitution for the deaf. We set out to build something totally unobtrusive—so unobtrusive that no one would even know you had it. To that end, we assimilated several advances in high-performance computing into a sound-to-touch sensory-substitution device worn under the shirt. Our Neosensory Vest captures the sound around you and maps it onto vibratory motors on the skin. People can *feel* the sonic world around them.

If it sounds strange that this could work, just note that this is all your inner ear does: it breaks up sound into different frequencies (from low to high), and then those data get shipped off to the brain for interpretation. In essence, we're just transferring the inner ear to the skin.

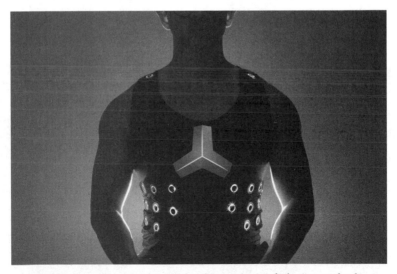

The Neosensory Vest. Sound is translated into patterns of vibration on the skin.

The skin is a mind-bogglingly sophisticated computational material, but we do not use it for much in modern life. It's the kind of material you'd pay great sums for if it were synthesized in a Silicon Valley factory, but currently this material hides beneath your clothing, almost entirely unemployed. However, you may wonder whether the skin has enough bandwidth to transmit all the information of sound. After all, the cochlea is an exquisitely specialized structure, a masterpiece of capturing and encoding sound. The skin, in contrast, is focused on other measures, and it has poor spatial resolution. Conveying an inner ear's worth of information to the skin would require several hundred vibrotactile motors—too many to fit on a person. But by compressing the speech information, we can use fewer than thirty motors. How? Compression is all about extracting the important information into the smallest description. Think about chatting on your cell phone: you speak and the other person hears your voice. But the signal representing your voice is not what is getting directly transmitted. Instead, the phone digitally *samples* your speech (takes a momentary record of it) eight thousand times each second. Algorithms then summarize the important elements from those thousands of measures, and the compressed signal is what gets sent to the cell phone tower. Leveraging these kinds of compression techniques, the Vest captures sounds and "plays" a compressed representation with multiple motors on the skin.[43]

Our first participant was a thirty-seven-year-old named Jonathan who was born profoundly deaf. We had Jonathan train with the Neosensory Vest for four days, two hours a day, learning from a set of thirty words. On the fifth day, Scott shielded his mouth (so his lips couldn't be read) and said the word "touch." Jonathan felt the complicated pattern of vibrations on his torso. And then he wrote the word "touch" on the dry-erase board. Scott then said a different word ("where"), and Jonathan wrote it on the board. Jonathan was able to translate the complicated pattern of vibrations into an understanding of the word that was spoken. He was not doing the decoding consciously, because the patterns were too complicated; instead, his brain was unlocking the patterns. When we switched to a new set of words, his performance stayed high, indicating that he was not simply memorizing but learning how to hear. In other words, if you have normal hearing, I can say a new word to you ("schmegegge") and you hear it perfectly well—not because you've memorized it, but because you've learned how to listen.

We've developed our technology into many different form factors, such as a chest strap for children. We have been testing it with a group of deaf children ranging from two to eight years old. Their parents send me video updates most days. At first it wasn't clear whether anything was happening. But then we noticed that the children would stop and attend when someone pinged a key on the piano.

The children also began to vocalize more, because for the first time they are closing a loop: they make a noise and immediately register it as a sensory input. Although you don't remember, this is how you trained to use your ears when you were a baby. You babbled, cooed, clapped your hands, banged the bars of your crib . . . and you got feedback into these strange sensors on the side of your head. That's how you deciphered the signals coming in: by correlating your own actions with their consequences. So imagine wearing the chest strap yourself. You speak aloud "the quick brown fox," and you *feel* it at the same time. Your brain learns to put the two together, understanding the strange vibratory language.[44] As we'll see a little later, the best way to predict the future is to create it.

Two children using the vibratory chest strap.

We have also made a wristband (called Buzz) that has only four motors. It's lower resolution, but more practical for many people's lives. One of our users, Philip, told us about his experience wearing Buzz at his work, where he accidentally left an air compressor running:

> I tend to leave it running and walk around the room, and then my co-workers say, "Hey, you forgot: you left the air on." But now . . . using Buzz, I *feel* that something is running and I see that it is the air compressor. And now I can remind *them* when they leave it running. They are always like, "Wait, how did you know?"

Philip reports he can tell when his dogs are barking, or the faucet is running, or the doorbell rings, or his wife calls his name (something she never used to do, but does routinely now). When I interviewed Philip after he'd worn the wristband for six months, I quizzed him carefully on his internal experience: Did it feel like buzzing on his wrist that he had to translate, or did it feel like direct perception? In other words, when a siren passed on the street, did he feel that there was a buzzing on his skin, which meant siren . . . or did it feel that there was an ambulance *out there*. He was very clear that it was the latter: "I perceive the sound *in my head*." In the same way that you have an immediate experience when seeing an acrobat (rather than tallying the photons hitting your eyes), or smelling cinnamon (rather than consciously translating molecular combinations on your mucosal membranes), Philip is hearing the world.

The idea of converting touch into sound is not new. In 1923, Robert Gault, a psychologist at Northwestern University, heard about a deaf and blind ten-year-old girl who claimed to be able to feel sound through her fingertips, as Helen Keller had done. Skeptical, he ran experiments. He stopped up her ears and wrapped her head in a woolen blanket (and verified on his graduate student that this prevented the ability to hear). She put her finger against the diaphragm of a "portophone" (a device for carrying a voice), and Gault sat in a closet and spoke through it. Her only ability to understand what he was saying was from vibrations on her fingertip. He reports,

> After each sentence or question was completed her blanket was raised and she repeated to the assistant what had been said with but a few unimportant variations. . . . I believe we have here a satisfactory demonstration that she interprets the human voice through vibrations against her fingers.

Gault mentions that his colleague had succeeded at communicating words through a thirteen-foot-long glass tube. A trained participant, with stopped-up ears, could put his palm against the end of the tube and identify words that were spoken into the other end. With these sorts of observations, researchers have attempted to make sound-to-touch devices, but in previous decades the machinery was too large and computationally weak to make for a practical device.

In the early 1930s, an educator at a school in Massachusetts developed a technique for two deafblind students. Being deaf, they needed a way to read the lips of speakers, but they were both blind as well, rendering that impossible. So their technique consisted of placing a hand over the face and neck of a person speaking. The thumb rested lightly on the lips and the fingers fanned out to cover the neck and cheek, and in this way they could feel the lips moving, the vocal cords vibrating, and even air coming out of the nostrils. Because the original pupils were named Tad and Oma, the technique became known as Tadoma. Thousands of deafblind children have been taught this method and have obtained proficiency at understanding language almost to the point of those with hearing.[45] The key thing to note for our purposes is that all the information is coming in through the sense of touch.

In the 1970s, the deaf inventor Dimitri Kanevsky came up with a two-channel vibrotactile device, one of which captures the envelope of low frequencies, and the other high. Two vibratory motors sit on the wrists. By the 1980s, inventions in Sweden and the United States were proliferating, demonstrating the power of a livewiring point of view. The problem was that all these devices were too large, with too few motors (typically just one) to make an impact.[46] It's only now that we're able to capitalize on advances in embedded computational power and price, signal processing, audio compression, energy storage, and the advent of inexpensive, wearable computing powerful enough to run sophisticated signal processing in real time.

Moreover, there are some advantages to this approach. Compare this with cochlear implants (like Michael Chorost's, whom we met at the beginning of the chapter), which cost about $100,000 to have implanted.[47] In contrast, modern technology can address hearing loss for some hundreds of dollars, which opens solutions for the whole globe. And implants require an invasive surgery, while a vibrating wristband is merely strapped on in the morning like a watch.[48]

There are many reasons to take advantage of the system of touch. For example, a little-known fact is that people with prosthetic legs have to do an enormous amount of work to learn how to walk with them. Given the high quality of prosthetics, why is walking so difficult? The answer is simply that you don't know *where* the prosthetic leg is. Your good leg is streaming an enormous amount of data to the brain, telling about the position of your leg, how much the knee is bent, how

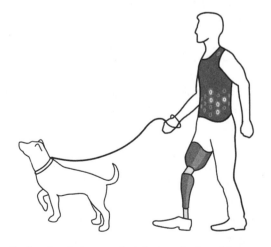

Feeding data from a prosthetic leg into the skin of the torso.

much pressure is on the ankle, the tilt and twist of the foot, and so on. But with the prosthetic leg, there's nothing but silence: the brain has no idea about the limb's position. So we attached pressure and angle sensors on the prosthetic leg and fed the data into the Neosensory Vest. As a result, a person can feel the position of the leg, much like a normal leg, and can rapidly learn how to walk again.

This same technique can be used for a person with a real leg that has lost sensation—as happens in Parkinson's disease and many other conditions. We use sensors in a sock to measure motion and pressure, and feed the data into the Buzz wristband. By this technique, a person understands where her foot is, whether her weight is on it, and whether the surface she's standing on is even.

Touch can also be used to address problems with balance. Remember the tongue display unit from Paul Bach-y-Rita? It can do more than vision. Consider Cheryl Schiltz, a rehabilitation counselor who lost her sense of balance after her vestibular system in her inner ear was poisoned by antibiotic treatments. She was unable to lead a normal life, perpetually being off balance and falling. She heard about a new development in which a person could wear a helmet outfitted with sensors that would indicate the tilt of the head.[49] The head orientation was fed to the tongue grid: when the head was straight up, the electrical stimulation was felt in the middle of the tongue grid; when the head tilted forward, the electrical signal moved toward the tip of the tongue; when the head tilted back, the stimulation moved toward the rear. Side-to-side tilts were encoded by left and right movement of the electrical signal. In this way, a person who had lost all sense of which way her head was oriented could *feel* the answer on her tongue.

Cheryl attempted a trial of the device with no small amount of skepticism. But there was an immediate effect: while she wore the helmet, her brain could understand the strangely routed information, and she could keep her head and body balanced. After a few sessions, she and the research team realized that there was a residual effect: if she wore the device for ten minutes, she would have about ten minutes of normal balance *after* she removed the helmet. Cheryl was so thrilled that she threw her arms around the researchers after these initial experiments.

But the news got better. Because her brain was rewiring itself from practice with the tongue grid, the residual benefits extended longer and longer after taking off the helmet. Her brain was figuring out how to take the whispers of signals—those that were undamaged—and strengthen them with the guidance of the helmet. After several months of use of the helmet, Cheryl was able to reduce her usage dramatically. The tongue grid had acted like neural training wheels, helping her to more clearly interpret whispers of residual signals and thereby build the necessary skills to outgrow the need for the device.

––––––––––

Sensory substitution opens new opportunities to compensate for sensory loss.[50] But that's just the first step of what can be done, and that leads us into the world beyond sensory substitution: sensory enhancement. What if you could take the current senses and make them better, wider, faster? What if you could not only fix broken senses but enhance existing ones?

ENHANCING THE PERIPHERALS

The goal of a therapeutic device is to return a deficit back to normal. But why stop there? Once a surgery is complete, or a device strapped on, why not crank things up so that one has talents beyond the species? This is not a theoretical point: there are many examples of superpowered sensory brains all around us.

––––––––––

In 2004, inspired by the promise of visual-to-auditory translation, a color-blind artist named Neil Harbisson attached an "eyeborg" to his head. The eyeborg is a simple device that analyzes a video stream and converts the colors to sounds. The sounds are delivered via bone conduction behind his ear.

So Neil hears colors. He can plant his face in front of any colored swatch and identify it.[51] "That's green," he'll say, or, "that's magenta."

Even better, the eyeborg's camera detects wavelengths of light *beyond* the normal spectrum; when translating from colors to sound, he can encode (and come to perceive in the environment) infrared and ultraviolet, the way that snakes and bees do.

Color	Sound Frequency
ultraviolet	over 717.6 Hz
violet	607.5 Hz
blue	573.9 Hz
cyan	551.2 Hz
green	478.4 Hz
yellow	462.0 Hz
orange	440.2 Hz
red	363.8 Hz
infrared	below 363.8 Hz

Left, *The color-blind artist Neil Harbisson wears the eyeborg.* Right, *His "sonochromatic scale" translates colors detected by the camera into sound frequencies. The inclusion of the highest and lowest frequencies allows the auditory system to overstep the normal limitations of the visual system.*

When it came time to update his passport photo, Neil insisted he didn't want to take off the eyeborg. It was a fundamental part of him, like a body part, he argued. The passport office ignored the plea: their policy disallowed electronics in an official photo. But then the passport office received support letters from his doctor, his friends, and his colleagues. A month later, the snap of his passport photo included the eyeborg, a success upon which Neil claims to be the first officially sanctioned cyborg.[52]

And with animals, researchers have taken this idea one step further: mice are color-blind . . . but not if you genetically engineer photoreceptors to give them color vision.[53] With an extra gene, mice can now detect and distinguish different colors. And the same can be done in squirrel monkeys, which normally have only two types of color receptors and are therefore red-green color-blind. But engineer in an extra color photoreceptor, as found in humans, and the monkeys enjoy human-level color experience.[54]

Or, more precisely, a *typical* human-level color experience. It turns out that a small fraction of human females have not just three kinds of color photoreceptors but four, and that means their brain figures out how to use all the information to make a new kind of sensory experience. They experience more unique colors and new mixtures of them.[55] When new peripherals are plugged in, useful information wins a voice in the functioning of the brain.

Sometimes enhancements happen accidentally. Many people go in for cataract surgery and have their lenses exchanged for a synthetic replacement. As it turns out, the lens naturally blocks ultraviolet light, but the replacement lens does not. So patients find themselves tapping into ranges of the electromagnetic spectrum they couldn't see before. One such patient, Alek Komarnitsky, is an engineer who had a lens replacement for his cataracts. And he now describes many objects as having a blue-violet glow that other people don't see.[56] He noticed this the day after his first cataract surgery when looking at his son's Colorado Rockies shorts. Everyone else sees the shorts as black, but he saw them with a faint blue-violet sheen. When he put a UV filter over his eye, he saw them like everyone else. When looking at a black light that's turned on, you will see nothing; Alek sees a bright purple glow. His new superpowers, which allow him to see past the normal spectrum of colors, grant him new kinds of experiences when looking at sunsets, gas stoves, and flowers.

At our Neosensory headquarters, engineer Mike Perrotta hooked up one of our wristbands to an infrared sensor. The first night I wore it, I was walking between buildings in the dark when I suddenly felt the wristband start to buzz. But why was there an infrared signal out here? I guessed there was an error in the code or the hardware, but nonetheless I followed my wrist in the direction of the signal, and the buzzing grew more intense. Finally I walked right up to an infrared camera surrounded by infrared LED lights. Normally, a nighttime camera like this remains invisible as it spies on us; with a wearable window into that part of the spectrum, it was immediately exposed.

A similar type of visual enhancement has been given to animals. In 2015, scientists Eric Thomson and Miguel Nicolelis plugged an infrared light detector straight into the brain of a rat. And the rat became able to use it. It could perform tests that required it to see, and utilize, infrared light in its choices. When a single detector was plugged into its somatosensory cortex, it took the rat forty days to learn the task. In a different experiment on another rat, they implanted three additional electrodes. That rat took only four days to master the task. Finally, they implanted the infrared detector directly into the visual cortex, and now it took a rat only a single day to conquer the task.

The infrared input is just another signal that can be made use of by the rat's brain. It doesn't matter how you get the information in there, as long as you get it in there. Importantly, the addition of the infrared sensor didn't take over or interfere with the normal function of the somatosensory cortex; the rat could still use its whiskers or paws to feel around. Instead, the new sense integrated itself smoothly. Eric Thomson, the postdoctoral fellow who led the studies, expressed his enthusiasm for what this all meant:

> I'm still pretty amazed. Yes, the brain is always hungry for new sources of information, but the fact that it actually absorbed this new, completely foreign type so quickly is auspicious for the field of neuroprosthetics.

Because of a long road of evolutionary particularities, we have two eyes placed on the front of our heads, giving us a visual angle on the world of about 180 degrees. In contrast, the compound eyes of houseflies give them almost 360 degrees of vision. So what if we could leverage modern technology to gain the joy of fly vision?

A group in France has done just that with FlyVIZ, a helmet that lets users see in 360 degrees. Their system consists of a helmet-mounted camera that scans the whole scene and compresses it to a display in front of the user's eyes.[57] The designers of the FlyVIZ note that users

Seeing in 360 degrees.

have a (nauseating) adjustment period upon first donning the helmet. But it's surprisingly brief: after fifteen minutes of wearing the headset, a user can grab an object held anywhere around him, dodge someone sneaking up on him, and sometimes catch a ball thrown to him from behind.

What if you could not only see in 360 degrees but also sense things that are normally invisible to you, like the location of several people around you in the dark?

Imagine that a team of private military contractors was dropped into a territory to hunt down hostile androids. Sound like an episode of HBO's *Westworld*? Indeed, as a scientific adviser for that show, I proposed our technology for this purpose. At the end of the first season, the "hosts" (the androids) foment a rebellion, so by season 2 an elite military team seeks to put down the insurrection. Wearing our Vests, the soldierly fighting men can *feel* the location of the hosts—in the dark, behind barriers, hiding where they least expect them. Off to the left in front of them two hundred yards, or just behind them, or on the other side of a wall. Although *Westworld* is set thirty years in the future, all of this is easy to accomplish with modern technology, and it expands human perception beyond the beautiful but limited eyeballs with which we come to the table.

I thought of the *Westworld* plot some months after we collaborated with Google to perform a very cool experiment with blindness. Several Google offices are outfitted with lidar (light radar), which is the spinning device you see on top of some auto-driving cars. In the office space, the lidar allows them to track the position of every moving object—in this case, humans moving around in the office.

We tapped into the lidar stream and hooked it up to the Vest. Then we brought in Alex, a blind young man. We strapped the Vest on him, and now—just like the soldiers in *Westworld*—he could feel the location of those moving around him. He could see in 360 degrees, going from blind to Jedi. And there was zero learning curve: he immediately got it.

Besides demonstrating the ease of extending our senses, Alex's experience corroborates the Potato Head model. Plug in a new data stream, and the brain will figure out how to use it. Alex's Vest, the FlyVIZ camera, and the rat with the infrared detector illustrate the irrelevance of tradition when it comes to biology. We can extend ourselves beyond the customs of our inherited genetics.

Expansions aren't limited to vision. Take hearing. Already, devices from hearing aids to our Buzz wristband can reach beyond the normal hearing scale. Why not expand into the ultrasonic range so that one can hear sounds only available to cats and bats? Or the infrasonic, hearing sounds with which elephants communicate?[58] As hearing technologies improve, there's no meaningful reason to limit inputs to the senses that happen to be typical for our species.

And consider smell. Remember the bloodhound dog, which can smell odors well beyond our comprehension? Consider building an array of molecular detectors and feeling different substances. Instead of needing a drug dog with its huge snout, you could directly experience that depth of odor detection yourself.

All of these projects open our windows onto the world, making some of the invisible visible. But beyond expanding senses to take in more of what they normally do, what if we could create entirely novel senses? What if you could directly perceive magnetic fields, or real-time data from Twitter? Because of the brain's remarkable flexibility, there is the possibility of leveraging these data streams directly into perception.

The principles we've learned so far allow us to think beyond sensory substitution, and beyond sensory enhancement, into the realm of sensory addition.[59]

CONJURING A NEW SENSORIUM

Todd Huffman is a biohacker. His hair is often dyed some primary color or another; his appearance is otherwise indistinguishable from a lumberjack. Some years ago, Todd ordered a small neodymium magnet in the mail. He sterilized the magnet, sterilized a surgical knife, sterilized his hand, and implanted the magnet in his fingers.

Now Todd feels magnetic fields. The magnet tugs when exposed to electromagnetic fields, and his nerves register this. Information normally invisible to humans is now streamed to his brain via the sensory pathways of his fingers.

His perceptual world expanded the first time he reached for a pan on his electric stove. The stove casts off a large magnetic field (because of the electricity running in a coil). He hadn't been aware of that tidbit of knowledge, but now he can *feel* it.

Reaching out, he can detect the electromagnetic bubble that comes off of a power cord transformer (like the one to your laptop). It's like touching an invisible bubble, one with a shape that he can assess by moving his hand around. The strength of the electromagnetic field is gauged by how powerfully the magnet moves inside his finger. Because different frequencies of magnetic fields affect how the magnet vibrates, he ascribes different qualities to different transformers—in words like "texture" or "color."

Another biohacker, Shannon Larratt, explained in an interview that he could feel the power running through cables and could therefore use his fingers to diagnose hardware issues without having to pull out a voltage meter. If his implants were removed, he says, he would feel blind.[60] A world is detectable that previously was not: palpable shapes live around microwave ovens, computer fans, speakers, and subway power transformers.

What if you could detect not only the magnetic field around objects but also the one around the planet? After all, animals do it. Turtles return to the same beaches on which they were hatched to lay their own eggs. Migrating birds wing each year from Greenland to Antarctica and then back again to the same spot. Pigeons who carry messages between kings or armies navigate with better precision than human messengers.

The Russian scientist Alexander von Middendorff wondered how these animals accomplished their magic, and in 1885 he correctly guessed that they might be using an internal compass, "like a magnetic needle for ships, those sailors of the air possess an inner magnetic feeling, which might be linked to the galvanic-magnetic flows."[61] In other words, they used the magnetic field of the planet to pilot their course.

Starting in 2005, scientists at Osnabrück University wondered if a wearable device could allow humans to tap into that signal. They built a belt called the feelSpace. The belt is ringed with vibratory motors, and the motor pointing to the north buzzes. As you turn your body, you always feel a buzzing in the direction of magnetic north.

At first, it feels like a pesky humming—but over time it becomes spatial information: a feeling that north is *there*.[62] Over several weeks, the belt changes how people navigate: their orientation improves, they develop new strategies, they gain a higher awareness of the relationship between different places. The environment feels more ordered. The layout of locations can be more easily remembered.

As one participant described the experience, "The orientation in the cities was interesting. After coming back, I could retrieve the relative orientation of all places, rooms and buildings, even if I did not pay attention while I was actually there."[63] Instead of thinking about moving through space as a sequence of cues, they thought about their routes from a global perspective. As another user put it: "It was different from mere tactile stimulation, because the belt mediated a spatial feeling. . . . I was intuitively aware of the direction of my home or of my office." In other words, his experience is not of sensory *substitution* (feeding vision or hearing through a different channel), nor is it sensory *enhancement* (making your sight or hearing better). Instead, it's a sensory *addition*. It's a new kind of human experience. The user goes on:

> During the first two weeks, I had to concentrate on it; afterwards, it was intuitive. I could even imagine the arrangement of places and rooms where I sometimes stay. Interestingly, when I take off the belt at night I still feel the vibration: When I turn to the other side, the vibration is moving too—this is a fascinating feeling![64]

Interestingly, after users take off the belt, they often report that they have a better sense of orientation for a while. The effect outlasts the tech. Just as we saw with the balance helmet, internal whispers of signals can get strengthened when an external device confirms them.[65]

What these humans experienced has been explored in deeper detail with rats. In 2015, scientists covered the eyes of rats and plugged digital compasses into the visual cortex. The rats rapidly figured out how to get themselves to the food in mazes, relying only on their head-direction signals.[66]

Whatever data the brain receives, it makes use of.

In 1938, an aviator named Douglas Corrigan revived a plane that came to be nicknamed the Spirit of $69.90. He then flew the plane from the United States to Dublin, Ireland. In those early days of aircraft, there were few navigation aids—generally just a compass combined with a length of string to indicate the direction of airflow relative to the plane. In a recounting of the event, *The Edwardsville Intelligencer* quoted a mechanic who described Corrigan as an aviator "who flies by the seat of his pants," and by most accounts this was the beginning of the expression in the English language. To "fly by the seat of one's pants" meant to steer by feeling the plane. After all, the body part that had the most contact with the plane was the pilot's rump, so that was the pathway by which information was transmitted to the pilot's brain. The pilot felt the plane's movements and reacted accordingly. If the aircraft slipped toward the lower wing during a turn, the pilot's buttocks would slide downhill. If the aircraft skidded toward the outside of the turn, a slight g-force pushed the pilot uphill. It wasn't until the end of World War I that a slip/skid indicator was invented—so early pilots were well practiced at estimating many factors (tilt, wind speed, external temperature, overall operation of the airplane) by paying close attention to their tactile senses, especially when they were flying through clouds or fog.

In this context, feeling data has a long history, and at Neosensory we're working to take it to the next level. Specifically, we're expanding the perception of drone pilots. The Vest streams five different measures from a quadcopter—pitch, yaw, roll, orientation, and heading—and that improves the pilot's ability to fly it. The pilot has essentially extended his skin *up there*, far away, where the drone is.

In case the romantic notion ever strikes you: airline pilots were *not* better when they flew by the seat of their pants, without instrumentation. Flights are made safer with a cockpit full of instruments, allowing a pilot to measure elements he otherwise can't access. For instance, a pilot can't tell from his derriere if he is flying level or in a banked turn.[67] Having rich tools is better than not having them; the problem pivots on getting their rich data into the brain. When you look at a modern aircraft, you'll see the cockpit is packed silly with instruments. The visual system has to crawl the scene one gauge at a time, which is a slow process. That's why it's interesting to rethink a state-of-the-art cockpit: instead of a pilot trying to read the whole thing visually, she *feels* it. A high-dimensional stream of data into the body tells the pilot what the plane is up to, all at once. Why does this have a good chance of working? Because the brain has great talent at reading high-dimensional data from the body. This is why, for example, you can balance on one foot: different muscle groups from your legs, torso, and arms are all streaming their data, and the brain summarizes the situation and rapidly sends out corrections.

So the difference between flying by the seat of the pants and flying by the skin of the torso has to do with the amount of data coming in. And because we live in a world bloated with information now, it's likely that we're going to have to transition from accessing big data to *experiencing* it more directly.

In that spirit, imagine feeling the state of a factory—with dozens of machines running at once—and you're plugged in to feel it. You become the factory. You have access to dozens of machines at once, feeling their production rates in relation to one another. You sense when things are running out of alignment and need attention or adjustment. I'm not talking about a machine breaking: that kind of problem is simple to hook up to an alert or an alarm. Instead, how can you understand how the machines are running in relation to one another? This approach to big data gives deeper patterns of insight.

Consider the wide-ranging applications for sensory expansion. Imagine feeding real-time patient data into surgeons' backs so that they don't have to look up at monitors during an operation. Or being able to feel the invisible states of one's own body—such as blood pressure, heart rate, and the state of the microbiome—thereby elevating unconscious signals into the realm of consciousness. Or imagine an astronaut feeling the health of the International Space Station. Instead of floating around and staring at monitors all the time, imagine patterns of touch summarizing the data from many different parts of the space station.

And let's take this one step further. At Neosensory, we've been exploring the concept of a *shared* perception. Picture a couple who feels each other's data: the partner's breathing rate, temperature, galvanic skin response, and so on. We can measure this data in one partner and feed it over the internet into a Buzz worn by the other partner. This has the potential to unlock a new depth of mutual understanding. Imagine your spouse calling from across the country to ask, "Are you okay? You feel stressed." This may prove a boon or a bane to relationships, but it opens new possibilities of pooled experience.

All of this has the chance to work because our incoming data streams fade into the background; we become aware of our senses only when our expectations are violated. Consider the feeling of your shoe on your right foot. You can attend to it and feel its presence. But normally, the data coming in from the skin of your foot live below your awareness. Only when you get a pebble in your shoe do you attend to the stream of information. And so it will go with data streaming in from space stations or spouses: you won't be aware of the sensations unless you focus on them, or until a surprise calls for your attention.

Imagine using patterns of vibration to feed your brain streams of information directly from the internet. What if you walked around with the Neosensory Vest and felt a data feed of neighboring weather cells from the surrounding two hundred miles? At some point you should be able to develop a direct perceptual experience of the weather patterns of the region, at a much larger scale than a human can normally experience. You can let your friends know whether it's going to rain, perhaps with far greater accuracy than meteorologists. That would be a new human experience, one that you can never get in the standard, tiny, limited human body you currently have.

Or imagine the Vest feeds you real-time stock data so that your brain can extract a sense of the complex, multifaceted movements of the world markets. The brain can do tremendous work extracting statistical patterns, even when you don't think you're paying attention. So by wearing the Vest around all day, and being generally aware of what's going on around you (news stories, emerging fashions on the street, the feel of the economy, and so on), you may be able to develop strong intuitions—better than the models—about where the market is going next. That too would be a very new kind of human experience.

You might ask, why not just use eyes or ears for this? Couldn't you hook up a stock trader with virtual reality (VR) goggles to view real-time charts from dozens of stocks? The problem is that vision is necessary for too many of our daily tasks. The stock trader needs her eyes to be able to find the cafeteria, see her boss coming, or read her email. Her skin, in contrast, is a high-bandwidth, unused information channel.

As a result of feeling the high-dimensional data, the stock trader might be able to perceive the big picture (*oil is about to crash*) long before she can pick out the individual variables (*Apple is going up, Exxon is sinking, and Walmart is holding steady*). How would this be possible? Think about the visual signals you get when you look at your dog in the yard. You don't say, "Well, I noticed there's a photon here, and a slightly dimmer one here, and a line of bright ones here." Instead, you perceive the big picture.

There are an unimaginable number of streams that could be fed from the internet. We've all heard of the Spidey sense: the tingling sensation by which Peter Parker detected trouble in the vicinity. Why not have a Tweety sense? Let's start with the proposition that Twitter has become the consciousness of the planet. It rides on a nervous system that has encircled the earth, and important ideas (and some non-important ones) trend above the noise floor and rise to the top. Not because of corporations who want to tell you their messages, but instead because an earthquake in Bangladesh, or the death of a celebrity, or a new discovery in space has captured the imagination of enough people around the world. The interests of the world ascend, just as do the most important issues in an animal's nervous system (*I'm hungry; someone's approaching; I need to find water*). On Twitter, the ideas that break the surface may or may not be important, but they represent at every moment what's on the mind of the planetary population.

At the TED conference in 2015, Scott Novich and I algorithmically tracked all the tweets with the hashtag "TED." On the fly, we aggregated the hundreds of tweets and pushed them through a sentiment analysis program. In this way, we could use a large dictionary of words to classify which tweets were positive ("awesome," "inspiring," and so on) and which were negative ("boring," "stupid," and so forth). The summary statistics were fed to the Vest in real time. I could *feel* the sentiment of the room and how it changed over time. It allowed me to have an experience of something larger than what an individual human can normally achieve: being tapped into the overall emotional state of hundreds of people, all at once. You can imagine a politician wanting to wear such a device while she addresses tens of thousands of people: she would get on-the-fly insight into which of her proclamations were coming off well and which were bombing.

If you want to think big, forget hashtags and go for natural language processing of all the trending tweets on the planet: imagine compressing a million tweets per second and feeding the abridgments through the Vest. You'd be plugged into the consciousness of the planet. You might be walking along and you'd suddenly detect a political scandal in Washington, or forest fires in Brazil, or a new skirmish in the Middle East. This would make you more worldly—in a sensory sense.

Now, I'm not suggesting that many people would *want* to tether themselves to the consciousness of the planet, but we could surely learn a lot from the proof of principle. It underscores that when we think about sensory addition, we have the freedom to think well beyond any normalcy that has come before.

Sometimes I've been asked why I would ever consider hooking up a human to these data streams rather than a computer. After all, couldn't a good artificial neural network do pattern recognition better than a human could?

Not necessarily. Computers can perform stirring feats of pattern recognition, but they have no particular skill at knowing what is important *to humans*. In fact, even humans often don't know in advance what is important to humans. That's why a human acting as a pattern recognizer is more wide lensed and more flexible than an artificial neural network could be. Take the stock market Vest as an example: when you stroll the streets of New York or Shanghai or Moscow, you take in subtle measures of what people are wearing, what products grab their attention, how optimistic or pessimistic they're feeling. You may not know in advance what you are looking for, but everything you see and hear feeds your internal model of the economy. When you additionally feel the Vest telling you the price fluctuations of individual stocks, the combination offers you a rich view. Because an artificial neural network is merely looking for patterns in the numbers fed into it, it is inherently limited by the choices of the programmers.

René Descartes spent a good deal of time wondering how he could know the *real* reality that surrounded him. After all, he knew our senses often fool us, and he knew that we often mistake our dreams for waking experiences. How could he know if an evil demon were systematically deceiving him, feeding him lies about the world that surrounded him? In the 1980s, the philosopher Hilary Putnam upgraded this question to "am I a brain in a vat?"[68] How would you know if scientists had removed your brain from your body, and were merely stimulating your cortex in the right ways to make you believe that you were experiencing the touch of a book, the temperature on your skin, the sight of your hands? In the 1990s, the question became "am I in the Matrix?" In modern times, it's "am I in a computer simulation?"

Such questions used to live in the philosophy classroom, but nowadays they're leaking over to the neuroscience laboratory. Recall that our normal experiences are nothing but sensory input. So signals plugged directly into the brain can accomplish exactly the same ends. After all, all the signals impinging on our sensors are converted to a common electrochemical currency, so we could circumvent the sensors and create the electrochemical signals directly. We can skip the middleman. Why should we push visual data through the ears or tongue if we can jack straight into the processor?

We already have the technology. The implanting of electrodes is typically done with a small number (one to a few dozen electrodes), and it's done deep in subcortical areas to address problems like tremor, depression, and addiction. To stimulate the cortex with a meaningful sensory message, one would need many more electrodes (probably hundreds of thousands) to stimulate rich patterns of activity.

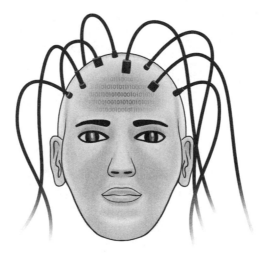

A new kind of augmented reality could be achieved by plugging new data streams directly into the cortex. The figure shows wires for illustration, but of course wireless is the future: you certainly don't want wires trailing behind you, subject to someone tripping on them like an unfortunate bride's train.

Several groups are working toward making this a reality. Neuroscientists at Stanford are working on a method to insert 100,000 electrodes into a monkey, which (if damage to the tissue is minimized) may tell us remarkable new things about the detailed characteristics of the networks. Several emerging companies, still in their infancy, hope to increase the speed of brain communication to the outside world by writing and reading neural data rapidly by means of direct plug-ins.

The problem is not theoretical but practical. When an electrode is placed into the brain, the tissue slowly tries to push it out, in the same way that the skin of your finger pushes out a splinter. That's the small problem. The bigger one is that neurosurgeons don't want to perform the operations, because there is always the risk of infection or death on the operating table. And beyond disease states (such as Parkinson's or severe depression), it's not clear that consumers will undergo an open-head surgery just for the joy of texting their friends more rapidly. An alternative might be to sneak electrodes into the tree of blood vessels that branches throughout the brain; however, the problem here is the possibility of damaging or blocking the vessels.

Nonetheless, there are possibilities on the horizon for getting information in and out of the brain at the cellular level, and these won't require electrode implantation. Within a decade or two, getting signals directly to the brain will be radically changed by massive miniaturization. Prospects include devices like neural dust: very tiny electrical devices that scatter around the surface of the brain, recording data, sending signals out to a receiver, and giving little zaps to the brain.[69]

There's also nanorobotics. Think 3-D printing with atomic precision. In this way, one can design and build complex molecules that are, in essence, microscopic robots. In theory, one can print a hundred billion of these robots, fit them into a little pill, and then swallow it. The nanorobots would, according to their imagined design, cross the blood-brain barrier, impregnate neurons, emit little signals whenever the neurons fire, and receive signals to force the neuron to become active. In this way, one could read and write to the billions of individual neurons in the brain. One could also leverage genetic approaches, building bio-nanorobots from proteins by encoding them in the DNA. There are many approaches to getting information into the brain, and in some decades we'll likely reach a stage in which each neuron can be read and controlled individually. At that point, our brains become our direct sensory enhancement devices, with no vests or wristbands required.

––––––––––

We've talked about ways of getting the data into the brain—whether from vibrations on the skin, shocks on the tongue, or direct activation of neurons—but this still leaves a major question: What would a new input *feel* like?

IMAGINING A NEW COLOR

Inside the vault of the skull, the brain has access only to electrical signals racing around among its specialized cells. It doesn't directly see or hear or touch anything. Whether the inputs represent air compression waves from a symphony, or patterns of light from a snow-covered statue, or molecules floating off a fresh apple pie, or the pain of a wasp sting—it's all represented by voltage spikes in neurons.

If we were to watch a patch of brain tissue with spikes dashing to and fro, and I were to ask whether we were watching the visual cortex or auditory cortex or somatosensory cortex, you couldn't tell me. *I* couldn't tell you. It all looks the same.

This leads to an unanswered question in neuroscience: Why does vision *feel* so different from smell? Or taste? Why is it that you would never confuse the beauty of a waving pine tree with the taste of feta cheese? Or the feeling of sandpaper on your fingertips with the smell of fresh espresso?[70]

One might imagine this has something to do with the way these areas are genetically built: the parts involved in hearing are different from the parts involved in touch. But upon closer examination, this hypothesis can't work. As we saw in this chapter, if you go blind, the part of the brain that we used to call the visual cortex gets taken over by touch and hearing. When one considers a rewired brain, it's difficult to insist there's anything fundamentally visual about the "visual" cortex.

So this leads to an alternative hypothesis: that the subjective experience of a sense—also known as its qualia—is determined by the structure of the data.[71] In other words, information coming from the two-dimensional sheet of the retina has a different structure than data coming from the one-dimensional signal on the eardrum, and it is different still from the multidimensional data from the fingertips. As a result, these data streams all feel different. A closely related hypothesis is that qualia are shaped by the way your motor outputs change the sensory inputs.[72] Visual data change as you send commands to the muscles around the eyes. The visual input changes in a learnable way: look to the left, and the blobby objects in the periphery sharpen into focus. As you move your eyes, the visual world changes, but that doesn't happen with sound. You have to physically turn your head for that. So that data stream has different contingencies. And touch is different still. We move our fingertips to objects, bringing them into contact and exploring them. Smell is a passive process that is magnified by a sniff. Taste blooms when you bring something to your mouth.

This suggests that we could feed a new data stream directly into the brain—such as data from a mobile robot, or your galvanic skin response, or long-wave infrared temperature data—and as long as there is a clear structure and a feedback loop with your own actions, the data should eventually give rise to new qualia. It will feel not like vision or hearing or touch or smell or taste—but like something entirely novel.

It certainly seems very difficult to imagine what such a new sense would be. And in fact, it is impossible to imagine it. To understand why, try imagining a new color. Go ahead. Squint your eyes and think hard. It seems as if this should be a simple task, but it's hopeless. In the same way you can't envision a new hue, you can't imagine a new sense.

Nonetheless, if your brain were consuming a real-time feed of data from a drone (pitch, yaw, roll, heading, and orientation), would the incoming data come to feel like *something*—in the way that photons or air compression waves do? And as a result, would the drone come to feel like a direct extension of your body? How about a more abstract input, such as the activity on a factory floor? Or Twitter feeds? Or the stock market? With the proper data streaming in, the prediction is that the brain will come to have a direct perceptual experience of manufacturing, or different hashtags, or the real-time economic movements of the planet. The qualia will develop over time; they are the brain's natural way of summarizing large amounts of data.

Is this a valid prediction, or pure fantasy? We are finally approaching the point, scientifically, where we will be able to put this to the test.

If the idea of learning a new sense seems foreign, just remember that you've done this yourself. Consider how babies learn how to use their ears by clapping their hands together or by babbling something and catching the feedback in their ears. At first the air compressions are just electrical activity in the brain; eventually they become experienced as sound. Such learning can be seen with people who are born deaf and eventually get cochlear implants as adults. At first, the experience of the cochlear implant is not like sound at all. One friend who got cochlear implants first described their effect as painless electrical shocks inside her head; she had no sensation that it had anything to do with sound. But after about a month it became "sound," albeit lousy like a tinny and distorted radio. Eventually she came to hear fairly well with them. This is the same process that happened to each of us when we were learning to use our ears; we simply don't remember it.

As another example, consider the joy of making eye contact with a newborn. The moment never lasts long, but it fills us with delight that we are among the first things seen by this new inhabitant of the world. But what if it turned out that you were not being *seen* at all? I suggest that vision is a skill that has to be developed. The brain takes quadrillions of spikes coming through the eyes and eventually learns to extract patterns, and patterns on top of those patterns, and patterns on top of those . . . and eventually the summary of all those patterns is what we call the experience of vision. The brain needs to *learn* how to see, just as it needs to learn how to control its arms and legs. Babies don't pop out knowing how to swing dance, nor do they have the subjective quality of vision. We had to learn our current sensory organs. And the same principles by which we did so can allow us to learn new sensory organs.

The fact that you can't imagine a new color is extraordinarily revealing. It illustrates for us the fence line of our qualia, beyond which we simply cannot walk. So if the ability to create new senses proves possible, a striking consequence is that we won't be able to *explain* the new sense to another person. For example, you have to experience purple to know what purple is. No amount of academic description will ever allow a color-blind person to understand purpleness. Similarly, make an attempt to explain vision to a friend born blind: you can try all you'd like, and your blind friend might even pretend to understand what you're talking about. But in the end it's a fruitless attempt. To understand vision requires experiencing vision.

Likewise, if you plug in an entirely new sense—and develop brand-new qualia—you won't be able to communicate it to others. First, we have no shared word for it. No one will understand. Language isn't all encompassing; it's only a way to tag things that we already share. It's a system of agreement about communal experiences. It's not that you couldn't attempt to articulate your new sense—it's simply that no one else has the foundations to understand it.

In a report of participants who had worn the feelSpace belt (the device that indicates magnetic north), the researchers wrote that the two users reported a change in perception and yet:

> articulating the perceptual quality they accessed and the qualitative experience arising from the different kind of spatial perception was hard. The observer got the impression that they lack concepts for what happens, such that they could only use metaphors and comparisons to come closer to an explanation.[73]

But was the problem the subjects' ability to articulate or the experimenters' ability to fathom? As the authors noted later, "It was much easier to talk about changes in perception between experimental subjects than to communicate it to naïve controls."

This is how things will go with the development of new senses. To understand them, we will have to feed in the data and learn the experience. So in some decades from now, if you're ever feeling lonely and misunderstood as you sit around with your new sense, the best solution will be to build a community of people who receive the same inputs. Then you can make up a new word for the internal experience—call it, say, "zetzenflabish." The word will make sense to your community, and no one outside it.

———————

With the right sort of data compression, what are the limits to the kinds of data we can take in? Could we add a sixth sense with a vibrating wristband and then a seventh with a direct plug-in? How about an eighth with a tongue grid and a ninth with a Vest? It's impossible at the moment to know what the limits might be. All we know is that the brain is gifted at sharing territory among different inputs; we saw earlier how smoothly it does so. And note that when rats had infrared sensors plugged into their somatosensory cortex, they gained the ability to see in that visual frequency band *without* losing the normal functions of feeling from the body. So it's possible the cortex doesn't have to engage in winner-take-all politics, but might instead be able to build a commune of senses much larger than expected. On the other hand, given the finite territory in the brain, is it possible that each added sense will reduce the resolution of the others, such that your new sensory powers will come at the cost of slightly blurrier vision and slightly worse hearing and slightly reduced sensation from the skin? Who knows? Answers about our limits remain pure speculation until they can be put to the test in the coming years.

However many senses we can add, there is another interesting question: Will those new senses carry emotional weight? For example, when you smell lemon pie fresh out of the oven, versus diarrhea on a sidewalk, you have different reactions: it's not zeros and ones on a screen; it's a full emotional response.

To understand this, ask *why* the pie smells good and the fecal matter smells bad. After all, the signals are not terribly different: in both cases, molecules diffuse through the air and bind to receptors in your nose. There is nothing inherent in a lemon-pie molecule or a fecal-matter molecule that makes it smell good or bad: these are simply floating chemical shapes, similar to the molecules that float off coffee, petunias, wet guinea pigs, cinnamon, fresh paint, moss on the side of a riverbank, or roasting chestnuts. All these shapes bind to a variety of smell receptors in the nose.

But we *like* the lemon-pie smell because the molecules predict the presence of a rich energy source. We have a bad emotion with the diarrhea because it is full of pathogens, and evolution doesn't want you, under any circumstances, to stick it in your mouth. This is akin to the way your visual system is confronted by a matrix of photons and might experience a wave of joy if the photons represent a grassy meadow, and shivers of disgust if it's a mutilated body. Or the way a pattern of spikes in the inner ear will be found delightful if it encodes a mellifluous tune consistent with your culture, and aversive if it's a baby screaming in pain. The emotions simply reflect the *meaning* of the data to you, in the context of your goals and evolutionary pressures. Many emotional examples have evolved on evolutionary timescales, but others come from experiences in your lifetime: consider the radio song that you love because it reminds you of a wonderful high school night, or the piece of clothing in your closet that gives you a bad feeling because it triggers the memory of getting dumped.

If the Potato Head model is correct, and the brain acts as a general-purpose computer, then this suggests that data coming in will eventually become associated with an emotional experience. Whatever the data stream, and however it gets there, it can carry passions.

Therefore, when feeding in a new stream of data from the internet, we may suddenly laugh with delight, cry with heartbreak, break out in goose bumps—depending on how the new data plug into our aims and ambitions. Imagine that you take on a new stream of stock market data. You suddenly get information that tech is tanking, and you're heavily invested in that sector. Will it feel bad? Not just cognitively bad, but emotionally aversive, like the smell of rotten meat or the sting of an ant bite? On the flip side, say that the information tells a brighter story: your investment has jumped by 6 percent. Will this feel good? Not just cognitively good, but emotionally pleasurable, like the sound of a baby's laughter or the taste of warm chocolate chip cookies?

If it seems strange that we could have these emotional reactions to new data streams, it bears remembering that *all* the meaning in our lives is simply built of data streams that carry importance in the context of our goals.

Finally, there's one more question worth asking before we close: Would having a new sense be overwhelming or stressful?

I don't think so. Consider a blind friend insisting to you that it must be stressful to have sight: *Just imagine having another data stream! You're catching a constant flood of billions of photons all the way from the distant horizon? You know what people are doing even when they're half a mile away? It must be nerve-racking to have that density of information streaming in all the time.*

If you are a sighted person, you know that vision is not particularly stressful. It typically lies somewhere between lovely and boring. You merge it into your reality effortlessly. Why? Because it's just another data stream, and incorporating data is what brains do.

ARE YOU READY FOR A NEW SENSATION?

In this chapter we looked at the creation of new senses. On an evolutionary timescale, if random genetic mutations manage to translate some source of information into electrical signals, the brain can treat them as plug and play, deciphering whatever information flows in. Plug eyes into a patch of cortex and it will become a visual cortex, plug ears into it and it will become an auditory cortex, plug skin in and it becomes somatosensory. This reveals one of the great tricks of nature: to tap into new energy sources from the world, it is not required to redesign the brain from scratch each time. Instead, she only needs to design new peripheral devices: light sensors, accelerometers, pressure sensors, heat pits, electroreception, magnetite, fingerlike noses, or whatever else she can dream up.

And whatever her creations can next dream up for themselves.

As seen with the cochlear implant of Michael Chorost or the retinal implant of Terry Byland, one can take advantage of the brain's flexibility by replacing an original peripheral device with an artificial one. The substitute device is not required to speak the native language of the brain, but instead can get by with a dialect that is close enough. The brain figures out how to use the data.

Taking the idea to the next step, we saw the power of sensory substitution. The brain's ability to rewire gives it tremendous flexibility: it dynamically reconfigures itself to absorb and interact with data. Thus we can use electrical grids to feed visual information via the tongue, vibratory motors to feed hearing via the skin, and cell phones to feed video streams via the ears. These devices can be used to endow the brain with new capacities, as we saw with sensory enhancement (extending the limits of a sense you already have) or addition (tapping into entirely new data streams). Such devices have moved rapidly from computer-laden cabled devices to sleek wearables, and this progress, more than any change in the fundamental science, will increase their usage and study.

As we'll expand on in the coming chapters, the brain reorganizes its circuitry to optimize its representation of the world. So when we feed in new, useful data opportunities, the brain wraps itself around them. This comes with two stipulations to which we'll return: the new data are best learned if they are tied to a user's goals and yoked to his own actions.

Given our current knowledge, there's no end to imagining the sensory expansions we'll build: vision in other parts of the electromagnetic spectrum, or ultrasonic hearing, or being plugged into the invisible states of your body's physiology. One might wonder whether this sort of technology will lead to a two-tiered society: the haves and have-nots. I think the risk of economic stratification is low, because these devices are inexpensive. Like the technology revolution that brought smartphones to the world (leapfrogging the personal computer revolution in most countries), sensory technology can be deployed worldwide at even lower costs than phones. This is not a technology confined to the rich.

Instead, I suspect that the future is much stranger than haves and have-nots: it will involve different flavors of having. Unlike the even blanketing of the planet by smartphones, we may possibly be heading into a future where different people will have different supersenses.

Imagine that you have a sense of oil futures, while your neighbor has trained up on the health of the space station, and your mother gardens by use of ultraviolet light perception. Might we be on the verge of a metaphorical speciation event, in which a species splits into multiple species? Who knows? In the best case, we might imagine a Hollywoodesque scenario in which a team of superheroes, each possessing a special power, comes together like pieces of a puzzle to conquer an archvillain.

The fact is that the future is hard to predict. Whatever the case, as we move toward the horizon, the only certainty is that we will increasingly choose our own plug-and-play peripheral devices. We are no longer a natural species who has to wait millions of years for Mother Nature's next sensory gift. Instead, like any good parent, Mother Nature has given us the cognitive capacity to go out and shape our own experience.

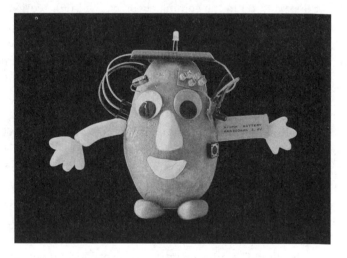

All the examples we've tackled so far involve *input* from the body's senses. What about the brain's other job, *output* to the limbs of the body? Is that also flexible? Could you embellish your body with more arms, mechanical legs, or a robot on the other side of the world controlled by your thoughts?

Glad you asked.

5

HOW TO GET A BETTER BODY

WILL THE REAL DOC OCK PLEASE RAISE HIS HANDS?

In *The Amazing Spider-Man* No. 3 (July 1963), a scientist named Otto Gunther Octavius plugs a device directly into his brain to control four extra robotic arms. The metal limbs, which operate as smoothly as his natural appendages, allow him to work safely with radioactive materials. Each of Dr. Octavius's limbs is able to operate independently—in the same way you might steer with one hand while changing the radio station with the other and depressing the gas pedal with your foot.

Unfortunately for Dr. Octavius, an explosion damages his brain and dooms him to a life of villainy. Guided by a new sense of immorality, he capitalizes on his extra arms to break into safes, scale buildings, and pioneer new methods of multiple-hand combat. Under the spell of his revised personality, he becomes known as Doctor Octopus, or Doc Ock.

When the comic book debuted in 1963, it was pure science fiction to imagine the brain could be attached directly to—and could effortlessly control—robotic limbs. But the idea has made a surprisingly rapid shift from fantasy to reality.

Earlier we witnessed the brain reorganize itself when a person lost a limb—as when Horatio Nelson's arm crossed paths with the musket ball. But that's only the *input* half of the story. On the *output* side, the cortex that drives the body (the motor map) adjusts itself as well. When the nervous system figures out that a previously existing limb is no longer available to control, the cortical territory devoted to it shrinks.[1] The brain reorganizes to match its new body plan.

Consider the case of a woman we'll call Laura, who lost her hand in a traumatic accident.[2] Her primary motor cortex began to shift over the course of weeks. The brain areas that controlled the neighboring arm muscles (like her biceps and triceps) slowly annexed the cortical territory that formerly operated her hand. We could phrase this another way: neurons that previously drove her hand were reassigned their job duty, now joining the team of the upper arm muscles. Laura's motor map was measured by shooting small magnetic pulses across her skull (a technique called transcranial magnetic stimulation) and noting which muscles twitched as a result. By this technique, researchers were able to watch the territory devoted to her upper arm muscles expand in a few weeks.

In later chapters, we'll learn how brains accomplish this trick, but for now we'll concentrate on *why* the motor system adjusts itself like this. The answer: the motor areas optimize themselves to drive the available machinery. And this principle opens the door to many possible body plans.

NO STANDARD BLUEPRINTS

A scan of the animal kingdom reveals a panoply of strange physiques, from the anteater to the star-nosed mole to the sloth to the dragonfish to the octopus to the platypus.

But here's something of a mystery: all the animals in the kingdom (including us) possess surprisingly similar genomes.

So how do creatures come to operate such wonderfully varied equipment—like prehensile tails, claws, larynxes, tentacles, whiskers, trunks, and wings? How do mountain goats get so good at leaping up rocks? How do owls get so good at plunging down upon mice? How do frogs get so good at hitting flies with their tongues?

To understand this, let's return to our Potato Head model of the brain, in which varied input devices can be attached. Exactly the same principle applies to output. In this view, Mother Nature has the freedom to experiment with outlandish plug-and-play motor devices. Whether fingers, flappers, or fins; whether two legs, four legs, or eight; whether hands or talons or wings—the fundamental principles of brain operation don't need to be redesigned each time. The motor system simply figures out how to drive the available machinery.

Wait a minute, you might say. If bodies are so easy to modify with tweaks of the genome, how come we don't find humans being born with strange varieties of body plans?

As it turns out, we do. For example, sometimes children are born with a tail,[3] demonstrating how easily genetic dominoes can be knocked over to produce larger structures.

Beyond tails, occasionally a human will come to the table with extra limbs. In Shanghai, for example, a boy named Jie-Jie was recently born with a fully formed third arm.[4] He had two full-sized left arms, one above the other.

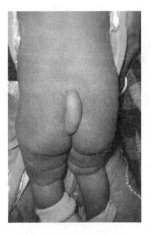

Children are sometimes born with tails, demonstrating how small genetic variations can generate large changes in body architecture.

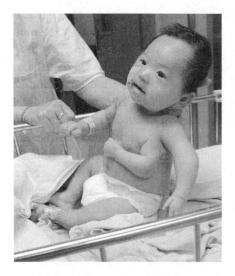

Jie-Jie was born with an extra arm.

This sort of thing sometimes happens because of a parasitic twin in the womb: one twin doesn't make it and gets absorbed into the body of the healthier twin. But this wasn't the case with Jie-Jie. His genetics simply dictated the growth of a third arm. A team of surgeons in China took several hours to remove the inner left arm, because both left arms were well developed. Often, an extra limb is shrunken and the choice of which to remove is easy. For Jie-Jie, both left arms had individual shoulder blades, making the operation challenging.

Tails and extra arms illustrate the way body plans can change unmistakably with small alterations of the genetics. And it goes without saying that this sort of genetic wobble happens in minor ways all around us: some people have longer arms, stubbier fingers, a big toe that is shorter than the second toe, wider hips, broader shoulders.

And although our nearest cousins, the chimpanzees, are nearly genetically identical to us, they possess many differences in the body plan; for starters, their biceps muscle has a higher insertion point, their hips are turned more outward, and their toes are longer. Perched in its dark throne room, the chimpanzee brain does not have a hard time figuring out how to drive a chimpanzee body to swing in trees

Matt Stutzman, the "Armless Archer."

and walk on knuckles, nor do human brains have a hard time figuring out how to compete in Ping-Pong and dance salsa. In both cases, the brain gracefully determines how best to drive the machinery in which it finds itself embedded.

To understand the power of this principle, consider Matt Stutzman, who was born without arms. He found himself attracted to archery, so he learned to manipulate a bow and arrow with his feet.

Moving fluidly, he notches the arrow into the string with his toes, then lifts the bow with his right foot. A strap holds the apparatus against his shoulder, allowing him to position the bow at eye level. He puts tension on the bow by drawing it forward with his foot, and when his aim is stably on target, he lets the arrow fly. Matt is not simply talented at archery, he is the best in the world: as of this writing, he holds the record for the longest accurate shot in archery. Probably not what his doctors would have predicted for a baby who came out of the womb armless. But they perhaps didn't realize how readily his brain would adapt its resources to solve problems in the outside world.

This sort of flexibility is seen across the animal kingdom. Consider Faith the dog. Faith was born without forelimbs, and she grew from puppyhood to be able to walk on her two hind legs, bipedally, like a human. Although we might have guessed that dog brains come hard-wired to drive standard dog bodies, Faith demonstrates how readily brains will navigate the world with whatever machinery they find themselves trapped in.

Brains adjust to the opportunities of the body. Faith the dog was compelled into bipedalism by an absence of front legs. Her motor systems adjusted to her body plan, allowing her to live a normal (if paparazzi-attracting) life.

Armless archers and bipedal dogs shine light on the fact that brains are not predefined for particular bodies, but instead adapt themselves to move, interact, and succeed. And this isn't simply about the body you're born in, but about whatever opportunities might come along. Take Sir Blake the bulldog, a California canine who has mastered skateboarding. Sir Blake leaps onto the skateboard, and with his front paw he scrapes at the ground to gain momentum. At the right moment, he sets his front paw onto the board and leans into the ride. He shifts his body weight to steer the board around obstacles, just as humans do. When he's done, he lets the board slow almost to a stop, and then he dismounts. Given the conspicuous absence of wheels in the evolutionary history of dogs, this underscores the adaptability of brains to steer new possibilities.

Although bulldogs evolved legs instead of wheels, Sir Blake
has little trouble adjusting to new methods for locomotion.

Or take another dog, Sugar, who took up surfboarding and has now been inducted into the International Surf Dog Walk of Fame. On second thought, forget Sugar—just be astonished that there *is* an International Surf Dog Walk of Fame. Canine brains aren't typically studied in the scientific context of how they hang ten on a longboard. But they can be. All they require is the opportunity, and their motor systems will figure it out.

Sugar won Best in Surf at the 2017 Surf Dog Surf-A-Thon
in Huntington Beach, California. In her frequent surfing matches,
her contenders range from golden retrievers to Pomeranians.

Sir Blake, Sugar, and their underdog competitors are shockingly good at riding the streets and the waves, and in some cases better than the creative species who invented these sports. How did these dogs get so good?

MOTOR BABBLING

A baby learns how to shape her mouth and her breath to produce language—not by genetics, nor by surfing Wikipedia, but instead by babbling. Sounds come out of her mouth, and her ears pick up on those sounds. Her brain can then compare how close her sound was with the utterances she's hearing from her mother or father. Helping things along, she earns positive reactions for some utterances and not for others. In this way, the constant feedback allows her to refine her speech until she mellifluously converses in English, Chinese, Bengali, Javanese, Amharic, Pemón, Chukchi, or any of the other seven thousand languages spoken around the globe.

In the same way, the brain learns how to steer its body by motor babbling.

Just observe that same baby in her crib. She bites her toes, slaps her forehead, tugs on her hair, bends her fingers, and so on, learning how her motor output corresponds to the sensory feedback she receives. In this way, she learns to understand the language of her body: how her outputs map onto the next inputs. By this technique, we eventually learn to walk, bring strawberries to our mouths, stay afloat in a pool, dangle on monkey bars, and master jumping jacks.

And even better, we use the same learning method to attach extensions to our bodies. Think about riding a bicycle, a machine that our genome presumably didn't see coming. Our brains originally shaped themselves in conditions of climbing trees, carrying food, fashioning tools, and walking great distances. But successfully riding a bicycle introduces a new set of challenges, such as carefully balancing the torso, modifying direction by moving the arms, and stopping suddenly by squeezing the hand. Despite the complexities, any seven-year-old can demonstrate that the extended body plan is easily added to the résumé of the motor cortex.

And this isn't just limited to typical bicycles. Consider Destin Sandlin, an engineer who was given a very odd bicycle by a friend: by dint of an elaborate gearing system, if Destin turned the handlebars to the left, the front wheel would turn right. And vice versa. Destin was fairly sure this wouldn't be difficult to master, because the concept was straightforward: steer the opposite direction that you want to go. But as it turned out, the bicycle was unbearably difficult to ride, because it required *un*learning the normal operation of a bicycle steering wheel. Training his motor cortex to master this new task was not as simple as having a cognitive understanding; after all, he *knew* how the bicycle worked. That didn't mean he could do the right thing with it.

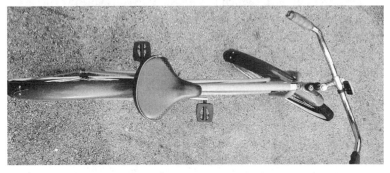

The bicycle with a reversed steering wheel operation.

But Destin began to get the hang of it. Each time he tried a move, he received feedback from the world (*you're falling to the left; you're crashing into a mailbox; you're swerving in front of a pickup truck*), and he used the feedback to adjust his next moves. After several weeks of daily practice, he was quite good at it. He mastered the strange bike in the same way he'd mastered a normal bike as a child: by motor babbling.

If you've ever driven in a country that has the steering wheel on the other side of the car, you know this sort of learning challenge. If you're an American driver in England, or vice versa, you swerve the wrong way many times on your way to getting the hang of things.

But you eventually get better, because your visual system looks at the consequences of each action and adjusts things accordingly. If all goes well, your nervous system completes its revisions before you plow into a hay bale.

––––––––––

It seems strange that we can learn to operate our body in different ways, given that you have only one motor cortex. Happily, the brain is extremely clever at using context to know what programs to run. It employs a schema (patterns to organize different categories of information) such that when you're on the bicycle, you transport yourself by moving your thighs in circles, but while you're jogging, you swing your arms and lift your feet to step over things in the road.

Here's an example by which I recently got to consciously experience my schema. The other day the rearview mirror of my truck broke off. I meant to fix it immediately, but I was busy writing this book, and so for a few weeks I drove around without it. The reason I finally fixed it was that something was driving me crazy: whenever I was in the seat of my vehicle, my eyes would make a ballistic movement up and to the right, and I'd find myself wondering why I was suddenly looking at the treetops on the side of the road. Obviously, my eyes kept shooting to the place the mirror was, and their intention was to see *behind* me. But of course, when I'm in my kitchen, office, or gym, I never shoot my eyes up and to the right to see behind me; I *only* do this when I'm in the driver's seat of my vehicle. The interesting part is how totally unconscious the schema has always been—assessing the affordances of my immediate surroundings and changing my motor functions accordingly. In the same way, when I'm jogging, I never squeeze my hand to come to a stop, and when I'm biking, I don't lift my foot to clear a stick on the ground.

Likewise, Destin's brain learned a new schema. When he finally mastered the trick bicycle, he found that he couldn't hop back onto a regular bike. But that was short-lived, and with a little practice he was able to master both. He can now hop onto either kind of bicycle, trick or normal, and his brain simply follows the pathways to operate his muscles in the current context.

Born into the world with some number of limbs and joints and actuators,
the Starfish robot figures out its own body and how to drive it.

Let's return to motor babbling. This is not only the way babies and
bicyclists learn how to move; it has also become a powerful new
approach in robotics. Consider a robot called the Starfish. It makes
an on-the-fly model of its own body, and therefore learns what it can
do. None of the typical programming is necessary; it learns its own
physique.[5]

The Starfish tries out a move, like an infant flailing a limb, and
assesses the consequences—in the robot's case, using gyroscopes to
see how the move tilted the central body. One stretch of the limbs can't
tell it what its body looks like and how it interacts with the world, but
the feedback narrows the space of possibilities. It now has a smaller
space of hypotheses about what it might look like. Now it's time for the
next move. Instead of doing a random movement, it chooses its next
move to best distinguish between the available remaining hypotheses.
By choosing each successive move in a manner to divide the possibil-

ity space in the right places, it develops an increasingly focused picture of its body.[6]

It uses motor babbling to learn how to operate its actuators, and that's why you can snap one of the legs off this robot and it will figure itself out again. It's like the Terminator after Sarah Connor has burned it and crushed its legs: it keeps going, operating with a different body plan but pursuing its goal nonetheless.

Building a babbling, self-exploring robot is more effective and more flexible than programming the robot to move in predetermined ways. And in the animal kingdom, nature has only some tens of thousands of genes with which to build a creature, so it cannot possibly preprogram all of the actions one might do in the world. Its only choice? Build a system that figures itself out.

And this trick is what allows the skateboarding and surfboarding dogs to master their craft. By babbling with their bodies, they try out various moves, postures, positions, and balances, and they assess the results. Does leaning to the left convey me along the wave, or dump me in the cold water? Does pushing with my rear leg while leaning keep the skateboard moving and my master squawking with delight, or does it cause a painful crash into a fire hydrant? The feedback allows the motor system to fine-tune millions of parameters and perform better the next time around. In this way, the organism builds a model of its body's interaction with the world. It comes to grasp its capacities and the consequences of its movements. It comes to know what the environment allows. By this constant loop of feedback, the baby, the athletic dogs, and the Starfish robot learn how to navigate their body plans. They nurture a feedback loop between the internal and the external worlds.

The loop of putting out actions and evaluating the feedback is the key to understanding not just motor babbling but also social babbling. Consider how you learned (and continue to learn) communication with other people. You constantly put social actions into the world, assess the feedback, and adjust. We rove the space of possibilities, trying out multiple personas when we're young: Is it better to be humorous in this situation, or cross your arms in a show of defiance, or cry and seek sympathy? We find traction with certain identities in particular situations, and we tend to stick with those until they require updates. And just like the human who at different times operates a mountain bike, ice skates, and a hang glider, we adopt different schemas for different social situations. Just as with motor feedback, we rely on social feedback. Does strong leadership work in this situation? Does a kind word get me what I need here? Does cracking an off-color joke succeed over dinner, while it gets me pilloried in the business meeting?

And this constant testing of the world may also be how we learn to *think*. From your brain's point of view, thinking is remarkably similar to motor movements. The neural storm of activity that causes your arm to lift is much like the storm that causes you to think about what you should say to your depressed friend, or where your other sock might have disappeared to, or what you're going to order for lunch. Thinking a thought is like moving a limb; in the same way that our brains drive a kick, a lunge, or a grasp, it may be that thinking moves concepts around in thought space. In other words, thinking is the act of pushing around concepts instead of coffee cups, notions instead of napkins. And this starts with the same sort of babbling: generating a thought and assessing the consequences. Some thoughts map well onto the world (*if I pull this cord, the lawn mower will start*), while others gain nothing (*what would happen if I fling my pancake across the table?*). Just like movements and speech, thoughts have to learn how to best operate in the world.

So back to Sir Blake, Sugar, and their canine companions. Beyond the pleasure they afford us when we watch them, they underscore a fundamental principle: if dog genetics yielded two legs instead of four, or wheels instead of paws, or a surfboard-like skeleton, the dog brain on the inside would not have to be redesigned. It would simply recalibrate.

Think of how effectively this strategy creates biodiversity. A livewired brain does not need to be swapped out for each genetic change to the body plan. It adjusts itself. And that's how evolution can so effectively shape animals to fit any habitat. Whether hooves or toes are appropriate to the environment, fins or forearms, or trunks or tails or talons, Mother Nature doesn't have to do anything extra to make the new animal operate correctly. Evolution really couldn't work any other way: it simply would not operate quickly enough unless body-plan changes were easy to deploy and brain changes followed without difficulty.

––––––––––

This massive flexibility is why we can so easily install ourselves into new bodies. Think back to Ellen Ripley, the protagonist in the original *Aliens* movie. In her climactic death match with the slimy alien, Ripley scrambles into an enormous robotic suit that allows her movements to be magnified into powerful metal arms and legs. At first she swings awkwardly, but after some practice she's able to land clanging punches on the alien's mucousy jaws. Ripley learns how to control her new, gargantuan body, and she does so thanks to her brain's capacity to adjust her relationship between her outputs (*swing my arm*) and her inputs (*where's that giant right arm now? Am I listing too far to the left?*). It's not difficult to learn these new associations, as demonstrated by forklift drivers, crane operators, and laparoscopic surgeons, all of whom get out of bed every morning to pilot strange new bodies. If Ellen Ripley's brain were a single-purpose device that could control only her standard-issue two-meter-tall human body, she would have been alien snack food.

Although the example is fictional, the principle behind it applies to the rollerblades, unicycles, wheelchairs, surfboards, Segways, skateboards, and hundreds of other devices we bind, buckle, or lash onto our natural bodies. The specifics of the devices' weight, joints, movements, and controllers—everything you can *do* with them—work their way into your brain circuitry.

In the earliest days of aviation, pilots used ropes and levers to make their flying machines extensions of their own bodies,[7] and the task for a modern pilot is of course no different: the pilot's brain builds a representation of the plane as a part of himself. And this happens with piano virtuosos, chain-saw lumberjacks, and drone pilots: their brains incorporate their tools as natural extensions to be controlled. In this way, a blind person's cane extends not just away from the body but into the brain circuitry.

So consider what this means for our near-term future as humans. Imagine you could control a robot at a distance, just with your brain activity. Unlike Ellen Ripley, you wouldn't even have to move: you would merely *think* a movement. When you want the robot to lift its arm, it would do so immediately. When you want it to squat, pirouette, or jump, it would heed your mental command without delay or mistake. Although this sounds like science fiction, it's already under way.

THE MOTOR CORTEX, MARSHMALLOWS, AND THE MOON

In early December 1995, Jean-Dominique Bauby was on top of the world: he was editor in chief at *Elle* magazine in Paris and revolved at the top of French social circles.

One afternoon, without warning, he suffered a massive stroke. He plunged instantly into a deep coma.

Twenty days later he awoke. He was mentally aware, he could see his surroundings, and he could understand what everyone was saying. But he couldn't move. He couldn't twitch his arms, his fingers, his face, his toes; he couldn't speak; he couldn't cry out. He discovered that his only available action was to blink his left eyelid. Other than that, he was locked in the frozen dungeon of his body.

Eventually, with the aid of a pair of persevering therapists, he was able to communicate, very slowly. Not by talking, but by winking his one functional eyelid. The therapist would slowly recite the letters of the alphabet in their order of frequency, and he would wink when she reached the correct letter. She would write down the letter and would start reciting the letters again. In this way, at the excruciating pace of two minutes per word, he could communicate. With mind-boggling patience, he wrote an entire book about the experience of living with locked-in syndrome. Its grace and eloquence contradicted the state of his body. He relayed the agony of being unable to interact with the outside world. He described, for example, the pain of seeing his assistant's purse lying half-open on the table: a hotel key, a metro ticket, a 100-franc note. The items reminded him of a life that for him was forever lost.

In March 1997, his book was published. *The Diving Bell and the Butterfly* sold 150,000 copies in the first week and became a number-one best seller across Europe. Two days after his book was published, Bauby died. Over the course of time, millions of readers dribbled tears all over the pages of his book. They appreciated, perhaps for the first time, the simple pleasure of having a functioning control center that successfully drives its enormous meat-robot and does so with such expertise that we are blessedly unaware of the massive operations running under the hood.

Why couldn't Bauby move? Under normal circumstances, when the brain decides to shift a limb, a pattern of neural activity sends the motor command down data cables in the spinal cord, out to the peripheral nerves, and there the electrical signals are converted into the release of chemicals (neurotransmitters), which cause the muscle to contract. But for Bauby, the signals never got out of the brain to make their long journey bodyward. His muscles never received the message.

Perhaps in the future we'll be able to fix damaged spinal cords, but at the moment that's not possible. So that leaves only one solution: What if we could have measured Bauby's brain spikes instead of his eye blinks? What if we could have eavesdropped on his neural circuits to figure out what they were trying to say to the muscles—and then bypassed the injury to make the actions happen in the outside world?

A year after Bauby's death, researchers at Emory University implanted a brain-computer interface into a locked-in patient named Johnny Ray, who lived long enough to control a computer cursor simply by imagining the movement.[8] His motor cortex was unable to get the signals through a damaged spinal cord, but the implant could listen and pass along the message to the computer.

By 2006, a paralyzed former football player named Matt Nagle was able to crudely open and close an artificial hand, control lights, open an email, play the video game *Pong,* and draw a circle on the screen.[9] Matt's powers emerged from a four-by-four-millimeter grid of almost

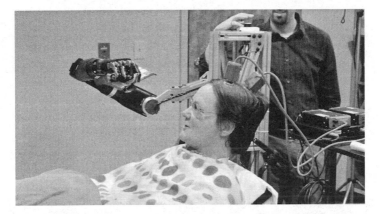

Controlling a robotic arm by imagining movements.

a hundred electrodes implanted directly into his motor cortex. He would imagine moving his muscles, which caused activity in his motor cortex, allowing researchers to detect the activity and crudely determine the intention.

The technology used with Johnny and Matt was makeshift and unpolished, but it proved the possibility. By 2011, neuroscientist Andrew Schwartz and his colleagues at the University of Pittsburgh built a prosthetic arm that was almost as sophisticated and lithe as a real arm. A woman named Jan Scheuermann had become paralyzed from a disorder called spinocerebellar degeneration, and she volunteered herself for neurosurgery to be able to control this arm.[10] Now, with signals recorded from her motor cortex, Jan can imagine making a movement with her arm, and the robotic arm moves. The robotic arm is partway across the room, but this makes no difference: through the bundle of wires attaching her brain to the machine, she can make it flowingly turn and grasp, essentially as well as she would have done with her own arm years ago. Normally, when you think about moving your arm, the signals travel from your motor cortex down your spinal cord, and then along your peripheral nerves and to your muscle fibers. With Jan, the signals recorded from the brain simply take a different route, racing along wires connected to motors instead of neurons connected to muscles. With time, Jan gets better at using the arm, in part

A thought-controlled robotic suit for people who are paralyzed.
Such neuroprosthetic devices are in early stages but edging toward the possible.

because of improving technology, in part because her brain is rewiring to understand how to best control its new limb—just as it would with a reversed bicycle, a surfboard, or Ellen Ripley's mech suit.

As Jan says, "I'd so much rather have my brain than my legs."[11] If you have the brain, you can build a new body, but not the other way around.

Currently, brain-machine interfaces are in active development to restore full-body movement to the paralyzed.[12] The Walk Again Project is an international collaboration that aims to help paralyzed people recapture mobility with a head-to-toe suit that moves according to brain commands. Just by thinking about movement, the way Jan does, a paralyzed person can be carried along. The idea is to surgically implant high-density arrays of microelectrodes into ten different areas of a volunteer's brain, allowing patients to channel their own brain activity to take command of sophisticated robotics.[13]

In 2016, researchers at the Feinstein Institute in New York took a slightly different approach. They eavesdropped on the motor system to know when it wanted to move muscles, but instead of feeding the information to a robotic arm or suit, they put the activation directly into the person's own muscles via an electrical stimulation system on the forearm.[14] A participant thinks about moving his arm, and the signals (passed through machine learning algorithms to learn how to best interpret the firestorm of neural activity) bypass the damaged spinal cord and jump instead to the muscle stimulator. The arm moves. Paralyzed participants are able to make different hand and wrist motions—grasping objects, manipulating them, and letting them go. They can even move individual fingers, allowing them to dial a phone, use a keyboard, or point toward the future.

———

If you redirect your outgoing brain signals to control a robotic arm, there's one shortcoming: the brain receives no sensory *feedback* from the fingertips. Whether you are grasping the egg too forcefully or too lightly can be assessed only by looking at it, and it's usually too late by the time you realize you weren't doing it correctly. It's like a baby verbally babbling while wearing earplugs.

The solution is to close the loop, and this can be done by zapping patterns of activity into the somatosensory cortex. When the robotic arm touches a target, feed a particular pattern of activity into the somatosensory areas—one that is equivalent to touch on the fingertips—and then the person feels as though her hand has touched a specific texture. When she touches a second target, she "feels" a different texture. In this way, she reaches out to touch the world and has a full sense of interaction with it. Her brain's flexibility eventually translates this into a full perception of the arm being her own. The brain learns to drive its body best when there's a closed loop of feedback: not just output, but input as well that verifies the interaction with the world. For example, when the baby bangs her arm against the bars of the crib, she feels it, sees it, and hears it.

Because most of the brain's learning takes place in this loop, it is no surprise that motor and sensory maps typically change hand in

Before tool use   After tool use

When a monkey uses a rake to retrieve distant objects,
the brain's representation of the body modifies to encompass the entire rake.
Oval shows the region in which a visible probe will cause a sensory neuron to fire.

hand with each other. For example, when monkeys are forced to reach for their food with a rake, their motor and somatosensory maps both reorganize to include the length of the tool. The rake literally becomes part of their body.[15] The motor and sensory systems are not fundamentally independent but instead yoked in an unbroken cycle of feedback.

We see that brain-machine interfaces can restore or replace damaged limbs. But could you use this same technology to add a limb?

In 2008, a monkey with two normal arms used its thoughts to control a third arm made of metal. By use of a tiny array of electrodes implanted in his brain, he controlled the robotic arm to pluck marshmallows and stick them in his mouth.[16] The monkey initially trained for this by moving a cursor on a screen toward a target, and he would get rewarded when he got it right. At first the monkey moved his own arms while doing the task. But something remarkable happened: he eventually stopped moving his arms, and the cursor continued to move on its own. His brain had rewired to separate out these tasks: some neurons corresponded to the real arm, some to the cursor on the screen. Eventually, the signals were able to control the robotic arm

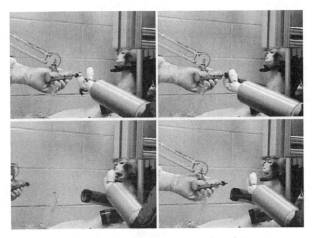

A monkey uses brain activity to control a robotic arm
and bring a marshmallow to his mouth.

for marshmallow gathering, all without any physical movement of the real arms. It had become a new limb.

Does it seem surprising that humans and monkeys can figure out how to move robotic arms with their thoughts? It shouldn't: it is the same process by which your brain learned to control your natural, fleshy limbs. As we've seen, your process as a baby was to flail your appendages around, bite your toes, grasp your crib bars, poke yourself in the eye, turn yourself over—for *years*—all toward fine-tuning control of your machinery. Your brain sent out commands, compared those with feedback from the world, and eventually learned your limbs' capabilities. Your skin-covered arm is no different from the clunky silver robotic arm of the monkey. It simply happens to be the standard operating equipment you're used to, so you often end up blind to its grandeur.

So while making a good robotic arm is a challenge for researchers, much of the operational work falls on the brain of the user. Because we didn't grow up with metal limbs, the movements of the souped-up tin can aren't intuitive. Our brains have to learn how to control the limb, just as Jan does. Half the work is on the side of the engineers, and the other half transpires in the neural forests of the user's brain.

The way the monkey learned to use the robotic arm independently of his real arms brings to mind Doc Ock, who controlled his robotic limbs even while performing more prosaic tasks with his fleshy appendages, such as pouring beakers of chemicals or steering a get-away car. The monkeys began to devote part of their brain real estate to the robotic arms, distinct from their natural limbs. They could divide resources and assign them to different appendages—flesh or metal.

For both Jan and the monkeys, the robotic arms were not directly connected to their torsos but instead connected by a bundle of wires. But if you could go wireless, the arm technically wouldn't need to be in the room. Could you control a robot on the other side of the world? Indeed, this has already been done.

Some years ago, the neuroscientist Miguel Nicolelis and his team at Duke University hooked up electrodes to a monkey, and the monkey controlled the walking patterns of a robot halfway across the globe, all in real time. As the monkey walked on a treadmill, the signals from his motor cortex were recorded, translated into zeros and ones, transmitted via the internet to a laboratory in Japan, and fed into a robot there. Like a metal doppelgänger, the five-foot-tall, two-hundred-pound robot walked as the monkey walked.

How did they get to this point? A good deal of legwork preceded this demonstration. First, Nicolelis's lab trained the rhesus monkey to walk on a treadmill. The researchers recorded from sensors on the monkey's legs to see how the muscles were moving, and they recorded from hundreds of brain cells to understand the translation from the neural activity to the muscle contractions. They turned the treadmill faster and slower to understand how brain activity correlated with the speed of the steps and the length of the strides.

Although no single neuron could tell them much, it became clear that neurons in different brain areas had particular timing relation-ships with one another, and this allowed the researchers to start unrav-eling the multi-muscle code underlying the deceptively complicated act of walking.[17]

With that research in place, they were now able to record from the monkey in North Carolina and send the real-time decoded motor commands to the robot in Kyoto. With the exception of small processing and transmission delays, the monkey and the robot strode in synchrony.

After they had demonstrated this proof of principle, the investigators at Duke stopped the treadmill. But as the monkey looked at its avatar on the screen, it *thought* about walking. And so the robot in Japan kept marching along. In the same way that Jan imagined movements and the arm executed them, the monkey's motor cortex continued to dream about the walk.

In the not-too-distant future, it seems inevitable that we will mind-control robots in factories, underwater, or on the surface of the moon, all from the comfort of our couches.[18] Our cortical maps, after extensive training, will have incorporated the robot's actuators and detectors: they will be our tele-limbs and our tele-senses. Our meaty bodies have evolved for the conditions of the oxygen-rich surface of this particular, idiosyncratic planet. But leveraging the brain's plasticity to build long-distance bodies is sure to change our main strategy for space exploration.

SELF-CONTROL

What consequence would an expansion of your body—say, a robotic arm or metal avatar across town—have for your conscious experience? The answer is that the robot will be perceived as part of you. It will be another limb. It will be an unusual limb, of course—because of the physical gap between you and it—but it will qualify as a new limb nonetheless. The only reason we're accustomed to connected limbs is that Mother Nature is a talented seamstress with muscle, sinew, and nerves—but she never worked out how to control distant limbs via Bluetooth.[19]

If extra limbs or tele-limbs seem exotic, recall that we have everyday experience with them. Just look in a mirror and move your arm. You'll see a distant object move in perfect synchrony with your motor commands. Although infants are at first confused by mirror images, they come to understand the reflections as themselves. Although they can't feel any direct sensation from the distant limbs, they can witness their *control* over them. And that's enough for those limbs to be annexed by selfhood.

This notion of the self is analogous to the Borg in *Star Trek,* who assimilate everything in their path into their singular identity—*except* for those things they cannot control, like the impossibly unpredictable Captain Picard.

The relationship of selfhood and predictability allows us to understand disorders such as asomatognosia, which translates to "not knowing one's body." In asomatognosia, damage to the right parietal lobe of the brain (say by a stroke or tumor) means a person is no longer able to control a limb. As a gobsmacking result, the patient will deny that the limb belongs to her and will sometimes insist that the limb belongs to someone else.[20] She will attribute the arm to, say, a dead friend, or a relative, or a phantasm, or a devil, or one of the medical professionals taking care of her. She will explain that her own, real limb was stolen or is simply missing. Variants of this disorder include construing the limb as an animal—perhaps a snake—with its own independent life force and intentions.

The manifestations can be varied and strange: a patient may feel indifferent toward her no-longer-self limb, or she may be delusional about it, coming up with strange fabrications to explain what has happened, such as "someone sewed this onto my body." Other patients may unsympathetically describe their limbs as something they dislike: "My leg is like a dead weight." In a more vicious version of this breakdown of selfhood, a patient may hate her alien limb, cursing at it and hitting it.[21]

There is no gold standard explanation for the disorder. However, you will have no trouble guessing my interpretation through the lens of this book: the brain can no longer control the limb, and so the limb falls from the brotherhood of the Self.

Sometimes these patients have a small window of lucidity in which they re-recognize their limb as their own. It doesn't last long. I hypothesize this may result when the arm *happens* to behave as they intended: accidental predictability. This might be a feeling of wanting to move the arm toward the chocolate bar on the table . . . and then the arm happens to move that way, leading the owner to take credit for the action. Given a person's lifelong experience of controlling his arm, it should come as no surprise that even a temporary impression of control can snap it back into alignment with the self, if only for a moment.

In the early 1970s, a different sort of loss of selfhood happened to the neurologist and writer Oliver Sacks.[22] While hiking in Norway, he was startled by the sight of a bull on his path. He scrambled away down a mountain path, and in his haste tumbled over a small cliff and detached the quadriceps muscle in his leg. After constructing a makeshift splint with his umbrella, he lurched down the mountain with his "utterly useless" leg until he was found by some reindeer hunters. Sacks then spent time in a hospital, delirious and confused. Because of the torn quadriceps, he could not move his leg—and he felt 100 percent certain that the leg was not his. At one point he thought that his leg was straight out in front of him, but then he discovered that it was hanging off the side of the bed. He was alarmed:

> I knew not my leg. It was utterly strange, not-mine, unfamiliar. I gazed upon it with absolute non-recognition. . . . The more I gazed at that cylinder of chalk, the more alien and incomprehensible it appeared to me. I could no longer feel it was "mine," as part of me. It seemed to bear no relation whatever to me. It was absolutely not-me—and yet, impossibly, it was attached to me—and even more impossibly, "continuous" with me. . . . There was absolutely no sensation whatever . . . it looked and felt uncannily alien—a lifeless replica attached to my body.

How do we understand Sacks's experience with his leg? Just like the Borg and Captain Picard, what you can control becomes the self, and what you cannot control has no relation to you. Because of his inability to make the leg follow his commands, Sacks had no feeling that it was *him*. Instead, it was just a foreign collection of billions of cells:

bone and skin and weird hair growing out of it. We would all look at our bodies this way if we could not drive them and had no sensation from them.

By the way, I suspect this sense of predictability is related to the way that a person you know deeply—say, a family member—becomes something like a part of yourself. Of course, humans are far too complex to predict perfectly, and the degree to which your spouse acts surprisingly is the extent to which he or she remains independent.

TOYS ARE US

One doesn't need prosthetics or brain surgery to try out new bodies. The developing field of avatar robotics allows a user to control a robot at a distance, seeing what it sees and feeling what it feels. Take the Shadow Hand, one of the most intricate artificial hands in existence. Each fingertip is equipped with sensors, which feed their data back into haptic gloves worn by the user. Sending data over the net, one can control a robotic hand in London from Silicon Valley.[23] Other groups are working on disaster-recovery avatars: robots that are sent in after earthquakes, terrorist attacks, or fires, to be piloted by drivers sitting somewhere safe. I haven't yet heard of people using strange-bodied avatars, but they certainly could: just as the brain learns skis, trampolines, or pogo sticks, it could learn to interface with a weird and wonderful avatar body.

Although avatar robotics will allow a small number of people to try out extended or strange bodies, it is frightfully expensive. Luckily, there's a better way to try out different body plans: inside virtual reality. Inside a simulated space you can make massive changes to your body plan instantly and inexpensively.

Imagine looking into a mirror in your VR world. You lift your arm, and you see your virtual avatar in the mirror raise an arm. You tilt your neck, and the avatar tilts its neck. Now imagine that your avatar has not your face but that of an Ethiopian woman, a Norwegian man, a Pakistani boy, or a Korean grandmother. For the reasons that we just saw about how the brain determines selfhood (*if I can control what it does, it is me*), it takes only a few minutes of prancing around in front of the mirror to convince yourself that you now inhabit a different body. You can then walk around in the virtual world as a different person, experiencing life through a modified identity. Self-identity is surprisingly flexible. Researchers have been studying in recent years how taking on the face of a different person can enhance empathy.[24]

But taking on a new face is just the beginning. In the late 1980s, because of a coding accident, the VR study of unusual bodies began. A scientist was inhabiting the avatar of a dockworker when a programmer accidentally made his arm enormous (about the size of a construction crane) by inserting too many zeros into the scaling factor. To everyone's surprise, the scientist was nonetheless able to figure out how he could operate accurately and efficiently with his mega-arm.[25]

This led thinkers to wonder what kinds of bodies could be occupied. The virtual reality pioneers Jaron Lanier and Ann Lasko rendered an experience in which people inhabited the bodies of eight-legged lobsters. Your two arms controlled the first two legs of the lobster, and the programmers tried out several (complicated) algorithms to control the other legs. It was tough work to control the eight legs of the lobster, but apparently some people were able to make it come to pass. Lanier coined the term "homuncular flexibility," pointing to the surprising elasticity of the brain's representation of its body.

Some years later, Stanford researcher Jeremy Bailenson and his colleagues set out to test homuncular flexibility more scientifically. They asked whether people could learn to accurately control a third arm in VR.[26] When you strap on the VR goggles and grasp the two controllers in your hand, you can see your own arms in the virtual space, and you see an additional arm as well, offering itself from the middle of your chest. The task is simple: touch a box as soon as it changes color. But there are lots of boxes, and to do well, you have to employ all three arms. The first two virtual arms are simply controlled by your own arms; the third arm is controlled by rotating your wrists. Within three minutes, users got it: they could accommodate to the new body plan, as measured by task performance.

There's no limit of physiques and frames to explore: imagine finding a virtual tail protruding from your tailbone, accurately controlled via your hip movement.[27] Or becoming the size of a golf ball, or the size of a building, or having six fingers, or becoming a housefly with wings. Or, like Doc Ock, becoming an octopus.

Marrying the flexibility of the brain to the burgeoning creativity of the VR design world, we're moving into an era in which our virtual identities will no longer be limited to the bodies that we happen to have evolved. Instead, we can now speed up evolution—from eons to hours. We can explore bodies that Mother Nature couldn't dream of, making virtual avatars neurally real.

And an interesting possibility is that changing your body might change your mind. One study concluded that college students who use avatars of senior citizens are more likely to put money in a savings account, that men who take on a female avatar behave in a more nurturing manner, and that people who witness their avatars exercising are more likely to subsequently exercise.[28] (In the world of fiction, such a body change was proposed to underlie Doc Ock's villainy: the idea was that the rewiring required to accommodate four extra appendages actually changed his thinking.[29]) In other words, who we are depends on how the whole brain is wired. Tweak the body and you may tweak the person.

As a real-life example, consider the metal smelter Nigel Ackland, who lost his forearm in an industrial accident. He was a physical and emotional wreck, but then he got a beautiful bionic arm installed.[30] His brain sends commands to his remaining nerves and muscles, and those signals get interpreted to move the hand smoothly in more than a dozen different motions. But here's the kicker: ask Nigel to turn his wrist. He'll hold up his arm and rotate his hand . . . and it keeps rotating, around and around on its axis. It continues to turn like a slowly spinning top as long as he wants it to. Nigel has a better body than you do, and by that I mean one with fewer constraints. When the bioengineers made this hand, they realized there's no particular benefit to sticking with the restricting ligaments and tendons that limit our movements. Presumably, Nigel can think thoughts we cannot. Like "keep my hand rotating." Or "install a lightbulb in one motion."

ONE BRAIN, INFINITE BODY PLANS

As we saw with Matt the archer or Faith the dog, brains adjust to drive whatever body they find themselves in. And like Jan's robotic arm and the monkey's marshmallow feeder, brains also figure out how to operate new hardware additions. Networks in the skull accomplish this trick by emitting motor commands (*lean to the left*), assessing the feedback (*skateboard tilts and wobbles*), and then adjusting parameters to climb the mountain of expertise.

Can the brain fit any world or any body plan? In some hundreds of years from now, we're likely to see human babies born on the moon or on Mars. They'll grow up with different gravity constraints. As a result, their bodies will likely develop differently, and they'll make use of different kinds of body extensions to get around. Neuroscientists in the distant future will study questions about their bodies and their brain development, and they'll also ask whether those babies will resultantly be different in other ways, such as their memory, cognition, or conscious experience.

Consider what it will mean to our industries when we learn the principles of livewiring. Think of the advantages if a car manufacturer could design an engine once and then drop it into any model of vehicle (lawn mower, three-wheeler, truck, spaceship) with the assumption that the engine will adjust itself to optimally drive that device. And imagine if the aftermarket could put new features on the car—such as fins or retractable legs—and let the vehicle figure out how to take advantage of them.

We're entering the bionic era where people can enjoy better and longer-lasting equipment than the meat-robots we came to the table with. When we are studied by unrecognizable descendants a million years from now, perhaps this moment in our history will be understood as the first time we stepped out from the slow haul of our development and began steering our body's future. Bionics will become increasingly common. When our great-grandchildren's legs stop functioning, they won't take that sitting down; when their arms are amputated, they won't accept handouts. Instead, they'll outfit the body with artificial limbs, knowing their brains will figure out how to control them. Paraplegics will dance in their thought-controlled exoskeletal suits.[31]

And beyond restoring lost function, they will extend their motor capabilities beyond our traditional biological constraints. In future centuries, the story of the eight-limbed Doc Ock will seem as quaint to readers as Jules Verne's science fiction daydream that man would be able to cross the Atlantic Ocean in less than a day. Our progeny won't have to limit themselves to the boundaries of their bodies; instead, they can extend across the universe according to whatever is under their control.

6

WHY MATTERING MATTERS

L ászló Polgár has three daughters. He loves chess and he loves his
daughters, so he launched a small experiment: he and his wife
homeschooled the girls on many subjects, and they trained them rig-
orously in chess. Daily, they hopped and skipped the assorted pieces
across the sixty-four squares.

By the time the eldest daughter, Susan, turned fifteen years old, she
became the top-ranked chess player in the world. In 1986, she quali-
fied for the Men's World Championship—a first-time achievement for
a female—and within five years she had earned the Men's Grandmaster title.

In 1989, in the middle of Susan's astounding accomplishments, her
fourteen-year-old middle sister, Sofia, achieved fame for her "Sack of
Rome"—her stunning victory at a tournament in Italy—which ranked
as one of the strongest performances ever by a fourteen-year-old. Sofia
went on to become an International Master and Woman Grandmaster.

And then there was the youngest sister, Judit, who is widely considered the best female chess player on record. She achieved Grandmaster status at the tender age of fifteen years and four months and
remains the only woman on the World Chess Federation's top 100 list.
For a while she held a position in the top ten.

What accounts for their success?

Their parents lived by the philosophy that geniuses are made, not born.[1] They trained the girls daily. They not only exposed them to chess; they fed them on chess. The girls received hugs, stern looks, approval, and attention based on their chess performance. As a result, their brains came to have a great deal of circuitry devoted to chess.

We've seen how the brain reorganizes in response to its inputs, but the fact is that not all information streaming into the pipelines is equally important. How brains adjust themselves has everything to do with what you're spending your time on.[2] If you decide to make a career change to ornithology, more of your neural resources will become devoted toward learning the subtle differences between birds (wing shape, breast coloration, beak size), while previously your neural representation of birds might have been something more crude (*is that a bird or an airplane?*).

THE MOTOR CORTICES OF PERLMAN VERSUS ASHKENAZY

A story is told about the violinist Itzhak Perlman. After one of his concerts, an admiring concertgoer said to him, "I would give my life to play like that."

To which Perlman replied, "I did."[3]

Each morning, Perlman drags himself out of bed at 5:15. After a shower and breakfast, he begins his four-and-a-half-hour morning practice. He takes a lunch and an exercise session, then launches his afternoon practice for another four and a half hours. He does this every day of the year, except for concert days, when he does only the morning practice session.

Brain circuitry comes to reflect what you do, and so the cortex of a highly trained musician morphs into something measurably different—in a way that you can see on brain imaging, even with an untrained eye. If you pay careful attention to a region of the motor cortex involved in hand movement, you'll find something amazing: musicians have a puckering on their cortex where nonmusicians do not, shaped roughly like the Greek letter omega (Ω).[4] The thousands

String player

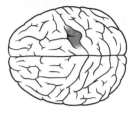

Pianist

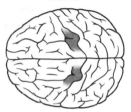

Differences between violin and piano players can be discerned
simply by looking at the motor cortex.

of hours of practice on the instrument physically molds the brains of
the musicians.

And the findings don't end there. Perlman the violinist and Vladimir Ashkenazy the pianist share a deep dedication to their craft,
countless hours of practice, and a grueling travel schedule—and yet
their brains look so different that you could easily discern which brain
belongs to whom. String players like Perlman show the omega sign
primarily in only one hemisphere, because the left fingers are doing
all the detailed work, while the right hand simply runs the bow over
the strings. In contrast, a pianist like Ashkenazy shows an omega sign
in both hemispheres, because both hands are performing meticulous
patterns on the ivories. Simply by eyeballing the motor cortex, you can
tell what kind of musician is in your scanner.

And we can read even more from the brain's reorganization: it represents not only that a hand is doing *less* or *more*, but sometimes *what*
it is doing. Say you get a job at an assembly line, and you're randomly
assigned to one of two jobs: either you put little marbles into jars, or

you turn the jar lid shut. Both jobs use your right hand, but the first requires fine use of the fingertips, while the second makes use of your wrist and forearm. If you're the jar filler, the cortical representation of your fingers will increase at the expense of your wrists and forearms. If you're the lid turner, the opposite happens.[5]

In this way, what you do over and over becomes reflected in the structure of the brain. And these changes involve much more than the motor cortex. For example, if you spend months learning to read Braille, the bit of your cortex that represents touch from the index finger will grow.[6] If you take up juggling as an adult, visual areas of your brain increase.[7] Brains reflect not simply the outside world but more specifically *your* outside world.

And this is what underlies getting good at something. Professional tennis players such as Serena and Venus Williams spend years of training so that the right moves will come automatically in the heat of the game: step, pivot, backhand, charge, fall back, aim, smash.[8] They train for thousands of hours to burn the moves down into the unconscious circuitry of the brain; if they tried to run a game just on high-level cognition, there's little chance they could win. Their victories emerge from crafting their brains into overtrained machinery.

You might have heard of the ten-thousand-hour rule, which suggests that you need to practice a skill for that many hours to become an expert—whether surfboarding, spelunking, or saxophone playing. Although the precise number of hours is impossible to quantify, the general idea is correct: you need massive amounts of repetition to dig the subway maps of the brain. Remember Destin Sandlin and the trick bike with the reversed steering? Although he cognitively knew how the bike worked, that proved insufficient to ride it. He needed to invest weeks of practice. Similarly, recall the monkeys who were forced to reach for their food with a rake. I told you that their body maps reorganized to include the length of the tool: the rake became part of their body plan.[9] What I didn't tell you at that time is that this reorganization works only when the monkey *actively* uses the rake. If the monkey passively holds it, there's no brain reorganization. The brain needs to practice repeatedly with the tool, not simply hold it. Hence the ten-thousand-hour rule.

The neural effects of intensive practice don't apply only to motor outputs, such as playing the violin, swinging a tennis racket, or manipulating a rake. They also apply to inputs. When medical students study for their final exams over the course of three months, the gray matter volume in their brains changes so much it can be seen on brain scans with the naked eye.[10] Similar changes transpire when adults learn how to read backward through a mirror.[11] And the areas of the brain involved in spatial navigation are visibly different in London taxi drivers from the rest of the population. In each hemisphere the taxi drivers have an enlarged region of the hippocampus, which is a region involved in internal maps of the outside world.[12] What you spend your time on changes your brain. You're more than what you eat; you become the information you digest.

And this is how the Polgár sisters were able to blossom into world-champion chess players. It is not because some gene codes for skill at chess; it is because they practiced over and over, chiseling pathways in their brains to encode the powers and patterns of knights, rooks, bishops, pawns, kings, and queens.

So brains come to reflect their world. But how?

FASHIONING THE LANDSCAPE

I recently saw an online meme: a picture of a human brain with a title that reads, "Hey, I think your phone just buzzed in your pocket." Across the bottom it reads, "Just kidding, your phone isn't even in your pocket, you moron."

The phantom cell phone vibration is a menace unique to the twenty-first century. It happens because of a momentary spasm or quivering or shaking or touch on your leg. As long as the frequency and duration of the feeling is vaguely similar to that made by your phone, your brain will decide on your behalf that someone interesting is trying to reach you. Thirty years ago, if you'd noticed a leg twitch, you would have interpreted the feeling as a fly landing on you, or a movement of your clothing, or someone accidentally brushing past you.

Why does your interpretation differ from one generation to the next? Because your phone now serves as the optimal explanation for a range of twitchy feelings.

To understand what's happening in the brain, think of a hilly landscape. For a raindrop to end up in a lake, it doesn't need to land directly in the water; instead, it only needs to hit the encircling hill-sides. Whether it lands on the northern slope or the southern, whether on the eastern incline or the western, it will slide down into the lake. Similarly, the feeling on your thigh doesn't have to be a buzzing phone. It can be a slight shift of your jeans, or a twitch of your thigh muscle, or an itch, or a graze past a sofa. As long as the feeling is close, the signals slide down the landscape to their conclusion: it's an important message that you can't wait to check. The landscape gets formed by what is important in your world.

Consider the way we interpret the sounds of language. It feels natural that you can understand the sounds of your native tongue, while foreign languages often have vexingly close sounds that you can't quite hear the difference between. But why? As it turns out, there is something different about the brains of the people who speak those languages.

But they weren't born that way, and neither were you.

If you look at the space of all the possible sounds humans can make with our mouths, it makes a relatively smooth continuum. Despite this, you learn from experience that specific sounds mean the same thing whether they're uttered by your father, your babysitter, or your teacher: your brain figures out that a drawn-out *eeeee* or a clipped *e* both belong to the *E* category. The same goes for a drawled *aeee* from your Texas friend, or an *oy* from your Australian friend. Experience teaches you that the speakers all mean the same sound, regardless of pronunciation, and so your neural networks carve out a land-scape in which all these sounds roll down the hillsides into the same interpretation.

In neighboring valleys you gather up sounds that are equivalent to *A*, or *I*, or *O*, and so on. With time, your landscape looks differ-ent from someone's who has grown up in another language and who needs to distinguish the smooth continuum of sounds differently than you do.

Take, for example, a baby born in Japan (call him Hayato) and a baby born in America (call him William). From their brains' point of view, there's nothing different about them. But in Osaka, Hayato hears Japanese all around him from day one. In Palo Alto, William hears the tones of English, where different sounds carry meaning. An example of what the two babies hear differently is the distinction between the *R* and *L* sounds. In English these carry information (right versus light, raw versus law), but in Japanese there is no such distinction between these two sounds. As a result, William's internal landscape builds a mountain range between his interpretation of *R* and *L,* such that the difference between these two sounds is perceptually clear. In Hayato's brain, the landscape develops into a valley, such that both *R* and *L* flow into the identical interpretation; as a result, Hayato cannot hear the difference between these two sounds.[13]

Obviously, the children's brains were not born this way: had William's pregnant mother moved to Osaka, and Hayato's pregnant mother to Palo Alto, the boys would have had no trouble in becoming fluent speakers and listeners in their new language. As opposed to a genetic issue, their neural landscapes were carved by what was relevant in their immediate environment.

The carving happens quite early, well before Hayato or William even learn to speak. This can be demonstrated by watching an infant's suckling behavior when there's a sudden change in sound. For example, make a continuous *R* sound, and then suddenly switch it to the *L* sound: *RRRRLLLL*. Infants, it turns out, will suck faster at the nipple when they detect a change in the sound—so at the age of six months, both Hayato and William will suck faster when the *R* swaps to *L*. But by twelve months, Hayato stops detecting the change. *R* and *L* sound the same to him, both sounds sliding down into the same valley. Hayato's brain has lost the ability to distinguish these, while William's brain, having passively listened to his parents speak tens of thousands of English words, has learned that there is information carried in the difference between the sounds. Hayato's brain, meanwhile, has picked up on other sound distinctions, which to William sound indistinguishable. Thus auditory system begins universally and then wires itself to maximize the distinctions unique to your language, depending where on the planet you happen to stick your head out of the womb.

Likewise, the buzzing phone isn't something you were born to detect; instead, its high relevance carves your neural landscape such that you have a broad catch-all for neighboring sensations. Like Hayato with his *R* and *L*, you combine your twitches, vibrations, and quiverings into a single interpretation of what just happened.

From what we've seen so far, we might think that repetitive practice or exposure is the key to molding the circuitry in your brain. But in fact, there is a deeper principle at work.

DOGGED

How many psychiatrists does it take to change a lightbulb?
Only one. But the lightbulb has to want to change.

Let's return to Faith the dog, the two-legged canine we met in the previous chapter. At that point, I told her story as though her brain had magically figured out her unusual body plan. But we can now dig a little deeper for a hidden bone. Was there something special about Faith? Could any dog have pulled this off? And if they could have, why don't all dogs walk bipedally?

Faith's rewritten maps were all about relevance to her life. Her brain was shaped by her goals. Faith needed to get to her food. That required a solution. It wasn't going to be the same one used by her four-legged siblings, nor was it to be found with drone delivery or DoorDash. She had to derive a novel solution. Her brain tried various strategies until it found one that worked: balancing on her two back legs and lurching forward step after step. This allowed her to get to what she needed, and after a while she became quite good at this method of locomotion. In the absence of finding an answer to her challenge, she would have starved and died. Her drive for survival allowed the flexible circuitry in her brain to try out many hypotheses and solve the problem, getting her to sustenance and shelter and the care of loved ones.

A brain's goals play a critical role in how and when it changes. For the Polgár sisters, Itzhak Perlman, or Vladimir Ashkenazy, achieving their expertise depended on a *desire* to achieve their expertise. Imagine for a moment that Serena and Venus Williams had a ne'er-do-well brother, Fred, and that their parents had put a tennis racket in his hands and forced him to go through all the years of practicing tennis. Imagine he found tennis repugnant. He never received good feedback from classmates about his performances, nor did he win any contests, nor did his elders lavish him with praise. The result of all that practice? Nothing. Fred's brain would show little reorganization. Although his body would be going through the motions, they would be misaligned with his internal incentives.

This is easily shown in the laboratory. Imagine an experiment in which someone taps out Morse code on your foot, while someone else, totally separately, plays a sequence of sounds. If you can win cash for decoding the messages on your foot, brain regions involved in touch to that part of your body (in the somatosensory cortex) will develop higher resolution. However, the regions involved in your hearing (the auditory cortex) will *not* change, even though that brain area is also receiving stimulation. Now imagine the reverse task: answering questions about subtle differences between the sounds earns the cash, while attending to the taps yields nothing. Now your auditory cortex will modify, but your somatosensory system won't.[14] The inputs from the world are exactly the same in both cases, but what changes depends on what is rewarded.

And this is why Fred Williams gets no better on the court: he derives no reward from it. In his brain, like yours, the present maps of neural territory reflect the strategies that have won positive feedback.

Such an understanding opens new paths for recovering from brain damage. Imagine that a friend suffers a stroke that damages part of her motor cortex, and as a result one of her arms becomes mostly paralyzed. After trying many times to use her weakened arm, she gets frustrated and uses her good arm to accomplish all of the necessary tasks in her daily routine. This is the typical scenario, and her weak arm becomes only weaker.

The lessons of livewiring offer a counterintuitive solution known as constraint therapy: strap down her *good* arm so that it cannot be used. This forces her to employ the weak arm. This simple method retrains the damaged cortex by forcing use of the bad arm—and by cleverly taking advantage of the neural mechanisms underlying desire and reward. After all, she has inherent motivation to get the sandwich to her mouth, to turn the key in her front door, to raise the cell phone to her ear, and to perform all the other actions that underlie a dignified, self-sufficient life. While constraint therapy starts off as frustrating, the approach proves to be the best medicine: it forces the brain to try new strategies, and reward locks in the methods that happen to work.

Remember the Silver Spring monkeys whose body maps changed? As it turns out, the idea of constraint therapy grew out of that research. Arm nerves had been impaired in each monkey, and the researcher, Edward Taub, began to wonder whether the monkeys stopped using the bad arm simply because the good arm was better at accomplishing the tasks. So Taub put this to the test by tying up the good arm in a sling, such that it couldn't be used at all. Now the monkey had a problem. One arm had a severed nerve, and the other was bound. If it wanted food, it really had only one choice: start using the weak arm. So it did just that. It seemed paradoxical that the solution to the monkey's malady was to make things worse, but that's precisely what helped the problem.[15]

So let's return to Faith the dog. Are all dogs able to walk on their back legs? Sure. But most dogs will never have the reason or motivation to attempt it, and certainly no cause to master it. And that's why Faith became famous: not because she's the only dog who *could* do it, but because she's the only one who made it happen. Similarly, recall the blind people using echolocation. It turns out that people with perfectly normal vision can learn to echolocate as well.[16] But most sighted people simply are insufficiently motivated to pour the hours into redefining their neural territory.

Reward is a powerful way to rewire the brain, but happily your brain doesn't require cookies or cash for each modification. More generally, change is tied to anything that is relevant to your goals. If you're in the far north and need to learn about ice fishing and different types of snow, that's what your brain will come to encode. In contrast, if you're equatorial and need to learn which snakes to avoid and which mushrooms to eat, your brain will devote its resources accordingly. Using relevance as its North Star, the brain flexibly picks up on important details. Its billions of neurons serve as a colossal canvas for painting the world we happen to find ourselves in, and with it we develop expertise in whatever has relevance to us, whether basketball, theater, badminton, Greek classics, cliff jumping, video gaming, line dancing, or wine making. When a task is roughly aligned with our larger goals, our brain circuitry comes to reflect it.

As an analogy, think of how governments continually self-design. In response to the attacks of September 11, 2001, the United States government altered its structure. It established the Department of Homeland Security, absorbing and restructuring twenty-two existing agencies. Similarly, the simmering Cold War initiated a large shift in 1947, spawning the Central Intelligence Agency.[17] In a thousand smaller ways, a government subtly mirrors the current aims of a nation and the events of its outside world. Budgets swell and shrink to echo priorities. When external threats loom, the military pocketbook expands; when peacetime follows, social initiatives gain. Like brains, nations respond to changing situations by shifting their resources and redrawing their organizational charts to meet the challenges they face.

ALLOWING THE REAL ESTATE TO CHANGE

How does the brain know when something important has happened and that it should change its wiring accordingly?

One strategy is to turn on plasticity when events in the world are correlated. That is, encode only those things that co-occur, such as seeing a cow and hearing a moo. In this way, related events become bound together in the tissue. Slow change is important here, because sometimes associations are spurious. For instance, you may see a cow but hear the bark of an unrelated dog. The brain would be ill-advised to permanently store every accidental co-occurance, so its solution is to change sluggishly, just a little at a time. In this way, it can encode only those things that commonly coincide. Real matches distinguish themselves from noise by occuring together over and over again.

But despite the wisdom of slow and steady change, extracting averages isn't the whole story. Consider one-trial learning, in which you touch a hot stove once and learn not to do it again. Emergency mechanisms exist to make sure that life- or limb-threatening events are permanently retained. But the story of one-trial learning goes deeper than that. Think back to when you were young and your aunt taught you a new word (*This is called a pomegranate*). You didn't need to learn this in an emergency situation, and nor did your aunt need to make the association a hundred times. She calmly told you once, and you got it. Why? Because it was salient to you. You loved your aunt, and you derived social benefit from knowing a new word and being able to ask for the fruit. This is one-trial learning not because of threat, but instead because of relevance.

Inside the brain, this relevance is expressed through widely reaching systems that release chemicals called neuromodulators.[18] By releasing with high specificity, these chemicals allow changes to occur only at specific places and times instead of all over at every moment.[19] An especially important chemical messenger is called acetylcholine. Neurons that release acetylcholine are driven by both reward and punishment. They're active when an animal is learning a task and needs to make changes, but not once the task is well established.[20]

The presence of acetylcholine at a particular brain area tells it to change, but it doesn't tell it *how* to change. In other words, when the cholinergic neurons (those that spit out acetylcholine) are active, they simply increase plasticity in the target areas. When they're inactive, there's little or no plasticity.[21]

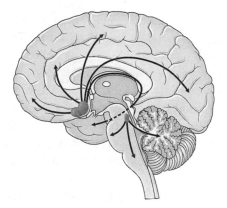

Acetylcholine has broad reach but tends to be released in very specific spots.
This allows rewiring in some areas and not others.

Here's an example. Imagine I play for you a particular note on the piano—say, F-sharp. The note triggers activity in your auditory cortex, but it doesn't change anything about how much territory is devoted to F-sharp. Why not? Because the note means nothing in particular to you. Now let's say every time I play the note, I give you a warm chocolate chip cookie. In this case the note accrues a meaning—and the territory devoted to F-sharp expands. Your brain assigns more ground to that frequency, because the presence of reward indicated that it must be important.

Now let's say I don't have any cookies available. So instead of handing you the treat, I play the F-sharp at the same moment I stimulate neurons in your head that release acetylcholine. The cortical representation for that tone expands, exactly as it did with the cookies.[22] Your brain allocates more terrain to that frequency, because the presence of the acetylcholine indicates that it must be important.

Acetylcholine broadcasts widely throughout the brain, and as a result it can trigger changes with any kind of relevant stimulus, whether a musical note, a texture, or a verbal accolade. It is a universal mechanism for saying *this is important—get better at detecting this.*[23] It marks relevance by increasing territory.

And changes in neural territory map onto your performance. This was originally demonstrated in studies with rats. Two groups were trained in a difficult task of grabbing sugar pellets through a small, high slot. In one group, the release of acetylcholine was blocked with drugs. For the normal rats, two weeks of practice led to an increase in their speed and skill and a correspondingly large increase in the brain region devoted to the forepaw movement. For the rats without the acetylcholine release, the cortical area didn't grow, and accuracy for reaching the sugar pellet never improved.[24] So the basis of behavioral improvement is not simply the repeated performance of a task; it also requires neuromodulatory systems to encode relevance. Without acetylcholine, the ten thousand hours is wasted time.

Recall Fred Williams, who (unlike Serena and Venus) hates tennis. Why doesn't his brain change, even after the same number of hours of practice? Because these neuromodulatory systems are not engaged. As he drills backhands over and over, he's like the rats grabbing the pellets without the acetylcholine.

Cholinergic neurons reach out widely across the brain, so when these neurons start chattering away, why doesn't that turn on plasticity everywhere they reach, causing widespread neural changes? The answer is that acetylcholine's release (and effect) is modulated by other neuromodulators. While acetylcholine turns on plasticity, other neurotransmitters (such as dopamine) are involved in the *direction* of change, encoding whether something was punishing or rewarding. Researchers all over the planet are still working to decipher the complex choreography of the neuromodulatory systems—but we know that collectively these chemical messengers allow reconfiguration in some areas while keeping the rest locked down.

London taxi drivers are famous for having to memorize the entire map of the streets of London. They train for many months on this task, and I mentioned before that there are physical changes in the structure of their brain as a result. The cabbies are able to pull off this staggering feat because the maps are relevant to them: this is their desired employment, which will pay for their home's mortgage or their child's tuition or their upcoming marriage or divorce.

But interestingly, since the study of the cabbies was first published in 2000, the need for such memorization has diminished. Now it's just as easy to have Google memorize all the streets of London, and more generally all the streets interlacing the planet.

As it turns out, current AI algorithms don't care about relevance: they memorize whatever we ask them to. This is a useful feature of AI, but it is also the reason AI is not particularly humanlike. AI simply doesn't care which problems are interesting or germane; instead, it memorizes whatever we feed it. Whether distinguishing a horse from a zebra in a billion photographs, or tracking flight data from every airport on the planet, it has no sense of importance except in a statistical sense. Contemporary AI could never, by itself, decide that it finds irresistible a particular sculpture by Michelangelo, or that it abhors the taste of bitter tea, or that it is aroused by signals of fertility. AI can dispatch ten thousand hours of intense practice in ten thousand nanoseconds, but it does not favor any zeros and ones over others. As a result, AI can accomplish impressive feats, but not the feat of being anything like a human.

THE BRAIN OF A DIGITAL NATIVE

How does the modifiability of the brain—and its relationship to relevance—bear on teaching our young? The traditional classroom consists of a teacher droning on, possibly reading from bulleted slides. This is suboptimal for brain changes, because the students are not engaged, and without engagement there is little to no plasticity. The information doesn't stick.

We're not the first generation to make this observation. The ancient Greeks noticed it. Lacking the tools of modern neuroscience but having a sharp eye, they defined several different levels of learning. The highest level—where the best learning occurs—is achieved when a student is invested, curious, interested. Through our modern lens, we would say that a particular formula of neurotransmitters is required for neural changes to take place, and that formula correlates with investment, curiosity, and interest.

The trick of inspiring curiosity is woven into several traditional forms of learning. For example, Jewish religious scholars study the Talmud by sitting in pairs and posing interesting questions to each other. (*Why does the author use this particular word rather than another? Why do these two authorities differ in their account?*) Everything is cast as a question, forcing the learning partner to engage instead of memorize. Although this is an ancient study structure, I recently stumbled on a website that poses "Talmudic questions" about microbial biology: "Given that spores are so effective in ensuring survival of bacteria, why don't all species make them?" "Do we know for sure that there are only three domains of life (Bacteria, Archaea, Eukarya)?" "How come peptides made enzymatically don't seem to get strung together to make a respectably sized protein?" The site has hundreds of such questions, coaxing active engagement in its readers instead of simply telling them answers. More generally, this is why joining a study group always helps: from calculus to history, it activates the brain's social mechanisms to motivate engagement.

In the 1980s, the author Isaac Asimov gave an interview with the television journalist Bill Moyers. Asimov saw the limits of the traditional education system with clear eyes:

> Today, what people call learning is forced on you. Everyone is forced to learn the same thing on the same day at the same speed in class. But everyone is different. For some, class goes too fast, for some too slow, for some in the wrong direction.[25]

Asimov held a vision of individualized education. Although he couldn't see the details, he was squinting into the future and anticipating the internet:

> Give everyone a chance . . . to follow up their own bent from the start, to find out about whatever they're interested in by looking

it up in their own homes, at their own speed, in their own time—
and everyone will enjoy learning.

It is through this lens of triggering interest that philanthropists such
as Bill and Melinda Gates aim to build adaptive learning. The idea is
to leverage software that quickly determines the state of knowledge of
each student and then instructs each on exactly what he needs to know
next. Like having a one-to-one student-teacher ratio, this approach
keeps each student at the right pace, meeting him where he is right
now with material that will captivate.

Like Asimov, Gates, and many others, I am a cyber-optimist on
the topic of education. Drilling down through blind mole holes of
Wikipedia, without any prespecified plan, may turn out to be a near-
optimal way to learn. The internet allows students to answer questions
as soon as they pop into their heads, delivering the solution in the
context of their curiosity. This is the powerful difference between *just
in case* information (learning a collection of facts just in case you ever
need to know them) and *just in time* information (receiving information the moment you seek the answer). Generally speaking, it's only in
the latter case that we find the right brew of neuromodulators present.
The Chinese have an expression: "An hour with a wise person is worth
more than one thousand books." This insight is the ancient equivalent
of what the internet offers: when the learner can actively direct her
own learning (by asking the wise person precisely the question she
wants to answer), the molecules of relevance and reward are present.
They allow the brain to reconfigure. Tossing facts at an unengaged
student is like throwing pebbles to dent a stone wall. It's like trying to
get Fred Williams to absorb tennis.

In this light, great opportunities are afforded by the gamification of
education. Adaptive software keeps students working at their point of
struggle, where finding the right answer is frustrating but achievable.
If the student can't get the answer, the questions stay at the same level;
when he gets the answer correct, the questions get harder. There's still
a role for the teacher: to teach foundational concepts and to guide
the path of learning. But fundamentally, given how brains adapt and
rewrite their wiring, a neuroscience-compatible classroom is one in

which students drill into the vast sphere of human knowledge by following the paths of their individual passions.

––––––––––

So the future of education looks favorable, but one question remains: Given that the brain becomes wired from experience, what are the neural consequences of growing up on screens? Are the brains of digital natives different from the brains of the generations before them?

It comes as a surprise to many people that there aren't more studies on this in neuroscience. Wouldn't our society want to understand the differences between the digital- and analog-raised brain?

Indeed we would, but the reason there are few studies is that it is inordinately difficult to perform meaningful science on this. Why? Because there's no good control group against which to compare a digital native's brain. You can't easily find another group of eighteen-year-olds who haven't grown up with the net. You might try to find some Amish teens in Pennsylvania, but there are dozens of other differences with that group, such as religious and cultural and educational beliefs. Where else can you find young people of the same age who don't have access to the internet? You might be able to turn up some impoverished children in rural China, or in a village in Central America, or in a desert in Northern Africa. But there are going to be other major differences between those children and the digital natives whom you intended to understand, including wealth, education, and diet. Perhaps you could compare millennials against the generation that came before them, such as their parents who did not grow up online—but who instead played street stickball and stuffed Twinkies in their mouths as they watched *The Brady Bunch*. But this is also problematic: between two generations there are innumerable differences of politics, nutrition, pollution, and cultural innovation, such that one could never be sure what brain differences should be attributed to.

So it is an intractable problem to pull off a well-controlled experiment about the effect of growing up with the internet. Nonetheless, I can tell you the root of my optimism. Never before have we had the entirety of humankind's knowledge in a rectangle in our pockets, with constant and immediate access. Some readers will remember trips to the library: we would pull out a volume of the *Encyclopaedia Britannica* (say, for the letter *H*) and we would flip through to the detail we wanted. The article had been written sometime in the previous decade or two. We'd hope it was sufficient, because otherwise we'd have to thumb through the card catalog and pray there was other material to be found at that library. And then our parents would take us back home because it was dinnertime.

In a surprisingly brief period, this has all changed. As a result, we've all seen the change in dinnertime debates: the winner has transitioned from being the most vociferous or persuasive to the person who can whip out the phone the fastest and google the fact in question. Discussions move rapidly now, leaping from one solved question to the next. And even when we're by ourselves, there is no end to the learning that goes on when we look up a Wikipedia page, which cascades us down the next link and then the next, such that six jumps later we are learning facts we didn't even know we didn't know.

The great advantage of this comes from a simple fact: all new ideas in your brain come from a mash-up of previously learned inputs, and today we get more new inputs than ever before.[26] Children now live in a time unparalleled in richness: our knowledge sphere has exploded in diameter, and as it grows it offers more doors for entry. Young minds have the opportunity to cross-link facts from completely different domains to generate ideas that previous eras couldn't have imagined. And this partially explains the exponential increase in human scholarship: we have faster communication and more mash-ups than ever before. It's not clear what all the social and political consequences of the internet will be, but from a neuroscience perspective it is poised to unlock a much richer level of education.

In earlier chapters we looked at brain changes resulting from modifications to the body plan, in terms of either sensors or limbs. In this chapter we turned to changes that result from practiced motor acts or rewarding sensory inputs. The larger principle that ties all those scenarios together is *relevance*. Your brain adjusts itself according to what you spend your time on, as long as those tasks have alignment with rewards or goals. For a person who goes blind, expanding the other senses takes on heightened relevance, and this is the deeper origin behind the changes that allow her visual cortex to be taken over. If a blind person passed her finger repeatedly over the bumps of Braille, but had no motivation to learn it, no rewiring would occur, because the right neuromodulators would not be present. Similarly, if adding a new telelimb to your body has relevance to you, your body will learn it—just as Faith the dog mastered a unique body plan.

Collectively, the principle of adjustment-from-relevance allows us to understand how animals constantly hurt themselves but keep on trucking, making the necessary brain adjustments to constantly progress toward their goals. Later we'll see how to leverage these principles to build new kinds of robots, ones that don't stop functioning when an axle breaks or part of the motherboard burns out or a screw wobbles loose.

But before we get there, we'll need to understand what drug withdrawal and broken hearts have in common, and why the notion of surprise matters for the brain's alchemy.

7

WHY LOVE KNOWS NOT ITS OWN DEPTH UNTIL THE HOUR OF SEPARATION

In the 1980s, tens of thousands of people began to notice something strange. When they looked at a floppy disk envelope with the black and white IBM logo emblazoned on the front, the letters appeared to have a red tint. The same thing happened when people looked at pages in a book: the page was shaded red. This happened only in the 1980s; people didn't perceive a red tint before or after. What was changing about brains during that window? To understand this, we first have to step back twenty-four hundred years.

A HORSE IN THE RIVER

The first recorded visual illusion was noted by the ever-observant Aristotle. He saw a horse stuck in a flowing river, and he fixatedly watched the rescue operation. When he finally looked away, it appeared that everything else—the rocks, the trees, the land—was flowing in the opposite direction of the river.

The easiest way to experience Aristotle's confusion and delight is to stare at a waterfall. After keeping your eyes locked on it for a bit, look over to the rocks to the side of the waterfall. The rocks appear to move upward.

The illusion has come to be known as the motion aftereffect. Why does it happen? The activity of particular neurons in your visual cortex represents downward motion, and the activity of other neurons represents upward motion. They're always locked in battle. Most of the time, the competition is evenly pitched, and they evenly inhibit each other. As a result, the world appears to be moving neither up nor down.

Given this, a popular explanation for the motion aftereffect is *fatigue*: by staring at the downward motion, you burn up a good deal of energy in your downward-coding neurons, and now their vigor is temporarily depleted. The battle therefore tips in favor of the upward-encoding neurons. As a result of this unbalanced activity, you perceive a net movement upward.

The fatigue hypothesis is appealing in its simplicity. But it's incorrect. After all, it can't explain some critical facts about the illusion. Imagine you watch the downward waterfall for a while and then close your eyes tightly—say, for three hours. When you reopen your eyes, you'll see the rocks crawling upward. That tells us it's not about a temporary energy depletion in neurons. There's something deeper going on.

The illusion comes about not because of passive fatigue but instead because of an active recalibration. Your system is exposed to continuous downward motion and, after a while, comes to assume this is the new normal. At first, the downward motion is dramatic information to the brain. After a while of staring, you receive no new information there. As far as your brain is concerned, this is the new reality: a world that flows more down than up. So your visual system carefully rebalances its expectations to mirror the world, to expect more downward than upward activity. Now, when you look away from the waterfall and toward the cliff side, the recalibrated setpoint becomes obvious, because now rocks and trees are flowing toward the sky. The setpoint (that is, what counts as standing still) has shifted.[1]

Why? The system always wants to set up a ground truth so that it can be better at detecting *change*. In this case, when your visual field is filled with the sight of the waterfall, your brain strives to subtract off the downward motion. All the descending flow is no longer informative, and so the circuitry adjusts itself to be maximally sensitive to new information.

You experience this sort of active recalibration all the time. When you get off a small boat, the land seems to be rocking for a while: it feels as though you are still on the water. What you are feeling is a negative aftereffect—the "negative image" of the water's motion.

And you have noticed this sort of illusion if you're a runner. Your body is used to sending motor commands (*run!*) to the legs, and the visual input flows by you as a result. But when you run on a treadmill at the gym, your brain doesn't get the flowing visual input. Instead, you're looking at a wall in front of you the whole time. When you step off, you experience the treadmill illusion: with each step you take toward the locker room, the scene seems to stream by at a faster pace. It looks as though you are moving forward more quickly than you really are.[2] Just as with Aristotle's horse or the waterfall or the post-boating wobbles, your brain is readjusting its expectations about the world—in this case, how the act of moving your legs should translate to the flow of the visual scene past your eyes.

As another example, take a look at the black and white lines below. Nothing special, right?

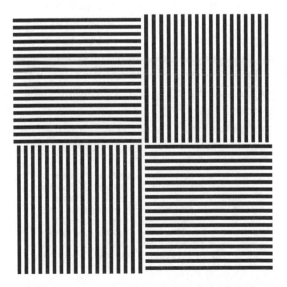

Now look at the red and green squares on the back of this book cover (or at eagleman.com/livewired): it is made up of green horizontal and red vertical lines. Stare at the colored lines for a bit: the red lines for a few seconds, then the green lines, then the red lines again, and then the green lines. Do this for about three minutes.

Now look back to the black and white lines above. You'll see that the spaces between the horizontal lines look reddish. And the spaces between the vertical lines look greenish.[3]

Why? Because when you stared at the colored figure, your brain realized that greenness had become tied to horizontal and redness to vertical, and so it adjusted to cancel out this strange feature of the world. When you looked at the black and white lines again, you experienced the aftereffect: the horizontal lines were being internally shifted toward the opposite color—red—and the vertical toward green. (Again, this has nothing to do with fatigue. In 1975, two researchers showed that if you stare at the red and green lines for fifteen minutes, the aftereffect can last three and a half months.[4])

This active recalibration of the world is why, in the 1980s, many people began to see text in books appearing to have a red tint. At that time, the population began to use computer monitors to do word processing. Unlike modern monitors, these early devices were able to display only one color, and so the text appeared as lines of green on a black background. People would stare at the horizontal green rows for hours at a time, and thus, when they picked up a book, the lines of text were shaded with the complementary color: red. Their brains were adjusting to a world of horizontal green lines, and their reality changed accordingly. The computer users also experienced this illusion when they looked at the IBM logo emblazoned on the front of the floppy disk sleeve: it looked tinged with red. Designers at IBM were

flummoxed about this; they had definitely not printed their black-and-white design with any red in it, and yet customers insisted they had.

How much motion is in the world, how stable the ground is, whether vision flows past us when we move our legs, whether lines are infused with color—none of this is decided in our genetics: it's calibrated by our experience.

MAKING INVISIBLE THE EXPECTED

If you look at a featureless scene that is entirely one color (say, yellow), the color will quickly drain away to neutral. Try it: take a yellow Ping-Pong ball and cut it exactly in half. Lay one hemisphere over each eye, and you'll see the whole world as an even blanket of yellow. But within a few moments there is no color at all. It is as though you are blind. Your visual system assumes the world has become yellower, and it adapts so that you'll be sensitive to other changes.

The scene doesn't have to be featureless to fade away like this. In 1804, the Swiss physician Ignaz Troxler noticed something stunning: if you stare at a central point in the middle of blobs, all the busyness in the periphery will eventually disappear. Keep your gaze firmly on the black dot in the middle for about ten seconds. Without looking at the surrounding blobs, note how they disappear into the background. Soon you're looking at a blank gray square.

Known as the Troxler effect, the illusion demonstrates that an unchanging stimulus in your peripheral vision will soon evaporate. Why does it happen? Because your visual system is always seeking motion and change. Something fixed quickly becomes invisible. Good information is expected to update; things that do not change are ignored by the system.

So what prevents, say, your kitchen or workplace from becoming Troxler-esque, with all the motionless features disappearing? First, most of the world is made up of hard edges, not blobs, and these are easier for your visual system to hang on to. But there's a deeper rea-

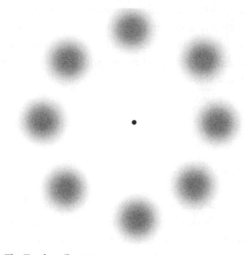

*The Troxler effect. Keep your eyes fixed on the center dot,
and all the blobs will fade away to nothingness.*

son. Although you are generally not aware of it, your eyes are constantly jumping and jiggling around. Observe your friend's eyes:
you'll notice that her eyeballs are making about three rapid jumps
every second of her waking life. If you watch even more closely, you'll
find that in between the big jumps, her eyes are constantly performing micro-jitters.[5] Is something wrong with her? No. These rapid
movements—both large and small—keep the retinal image fresh.
Totally unconsciously, her eyes labor to maintain a constantly changing image. Why do they bother? Because any image that remains perfectly fixed in one position on the retina will become invisible.

Here's how to prove this to yourself. If you wear contact lenses, take
a marker and draw a small shape on the front of your contact, right
in the middle. When you put the contact back in place on your eye,
you'll see something there—but it won't last long. It rapidly fades to
invisibility.[6] This phenomenon underscores the fundamental fact that
brains care about change. Just as in the Troxler effect, features that
don't change yield little information about the world. All the important information comes from things in flux.

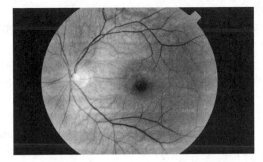

The retina is blanketed by a meshwork of blood vessels. Because these sit between the world and our photoreceptors, we should see them superimposed on the visual scene. But the vessels' pattern is stable, and hence carries no new information— and so our visual systems learn to ignore them completely.

If you don't have contact lenses, don't worry: you're already performing a similar experiment without knowing it. You have blood vessels that sit on top of your retina, at the back of your eye. This weblike retinal vasculature *should* be seen superimposed on top of everything you look at, because it's smack in front of your photoreceptors. However, it's totally invisible to your perception. Just like the drawing on the contact lens, the vasculature is fixed in position with respect to the retina. No matter how many movements your eyes make, they are never "refreshing" the image of these blood vessels. Even though the vessels interpose themselves between you and the world, they perceptually disappear like a magic trick.

You might have noticed a flash of these vessels when the ophthalmologist shines a penlight in your eyes.[7] In this situation, the beam of light can cause the vessels to cast a shadow at an unusual angle, and your visual system suddenly takes notice. Something unexpected has just occurred at the retina, and that is the only time you witness this massive network obstructing your view. (If you haven't seen this before, put down the book, go into a dark room, and shine a light in your eye from an angle. You'll see the blood vessels appear in front of you. Your visual system will adapt fairly quickly, so the trick is to keep moving the light to different angles to maintain the image.)

The strategy of ignoring the unchanging keeps the system poised to detect anything that moves or shifts or transforms. At the extreme, this is how reptile visual systems work: they can't see you if you stand still, because they only register change. They don't bother with position. And such a system is perfectly sufficient: reptiles have been surviving and thriving for tens of millions of years.

So let's return to the waterfall illusion. Why doesn't your visual system shift so much that the waterfall is perceived as standing still? First, there may be limits to the recalibration:[8] it simply cannot recalibrate enough to subtract off the massive motion of the falls. But there's another possibility: you haven't watched the waterfall for long enough, and if you did, it would eventually recalibrate all the way. How long might that take? Two months of staring? Two years? In theory, if you were to watch for enough time, then the short-term changes in your visual system would eventually lead to longer-lasting changes, leading eventually to changes at the deepest levels of the system (we'll return to this cascade of changes in chapter 10). Ever-present background motion would become invisible to us.

And this leads to a wild—but logically sound—speculation: Are parts of the world invisible to us that should be obvious? Imagine there were something like a cosmic rain that had existed your entire life. It would be completely invisible to you. Having never seen otherwise, your industrious visual system would have set the downward motion as its zero point. If the cosmic rain suddenly stopped, it would seem as though the whole world were suddenly moving upward. We would believe that something had just appeared—ascending rain— even though the real rain had just ended. And this situation could happen in any sensory channel: Imagine the beep-beep-beep of a cosmic alarm clock with no snooze button. All the time, all throughout the cosmos: beep-beep-beep. If it were totally regular, you wouldn't hear it at all, because your brain would have adapted to it. If the cosmic alarm clock suddenly ceased, everyone would suddenly hear a great beep-beep-beep, but we would have no idea that we were experiencing the aftereffect, with the "external" sound totally inside our heads.[9] Successful adaptation makes regularities invisible.

THE DIFFERENCE BETWEEN WHAT YOU THOUGHT WOULD HAPPEN
AND WHAT ACTUALLY HAPPENED

We've been talking about these illusions as results of adaptation, but there's another way to look at this: as *prediction*. If you subtract away the downward motion of the waterfall, or the rocking of the boat, or the drawing on your contact lens, this is equivalent to predicting its continued existence. When brain circuitry adjusts, it's making a guess about what the world is likely to be in the next moment. It stops talking about news that is expected to continue. Think about your retinal blood vessels. They are perceptually invisible because your visual system predicts them away: it knows they are going to be there, so it ignores them. Only if those expectations are violated (as when a light shines in from a strange angle) does your brain spend energy on representing the data.

Your brain doesn't want to pay the energy cost of spiking neurons, so the goal is to reconfigure the network to waste as little power as possible. If a pattern streams in that is predictable—or even partially guessable—the system saves energy by structuring itself around that input so as not to be surprised by it. A quieter nervous system means fewer violations of expectations: things in the outside world are going approximately as forecast. In other words, an energy-conscious brain wants to predict away everything possible so that it can save its energy for just representing the unexpected. Silence is golden. While many neuroscientists think of activity in neurons as the representation of things in the world, it could turn out that the truth is exactly the opposite: spikes are the unpredicted, energy-expensive part. The representation of something totally expected would be nothing but a hush falling over the neuronal forest.

The system makes adjustments only when it gets surprised. If your brain believes that all bricks weigh the same amount, and then you attempt to pick up a brick made of lead, the violation of your expectation causes cascades of changes to deal with this new turn of events. In contrast, when everything is successfully predicted away, there is no need to change anything. For these reasons, when you first look at the Troxler picture, you notice the blobs, and when you first put in the contact lens, you detect the drawn-on shape. But after a short time your brain adjusts itself. It is no longer surprised.

As another example of the brain predicting things away, consider this. When people first experience the Neosensory wristband (which converts sound to patterns of vibration on the skin), they typically say with surprise, "Whoa, it's picking up on my own voice!" They are always startled by that: it seems they shouldn't be registering their own speech. But of course your ears pick up on your voice all the time. It's typically the loudest voice in your conversations, because your own mouth is the closest one to your ears. However, because you can perfectly predict your own vocalizations, you hardly "hear" your own voice. Wearers of the wristband are also struck by the volume of other predictable sounds they normally pay no attention to (because they create them): flushing the toilet, closing the door behind them, their own footsteps. It's not that your auditory system doesn't register these sounds, but instead that you actively predict them away. This becomes obvious when wearing the wristband: you can't believe how loud these events are, because your brain has not yet learned to predict the signals coming up your arm.

Your brain actively recalibrates, because that allows it to burn less energy. But there is an even deeper principle at work here. In the darkness of the skull, your brain is striving to build an internal model of the outside world.

When you walk around your home, you pay little heed to the environment, because you already have a good model of it. In contrast, when you're driving in a foreign city, trying to find your way to a particular restaurant, you are forced to look around at everything— the street signs, the store names, the building numbers—because you don't already have a good model of what to expect.

So how do you build up a good internal model? What is the neural technology that allows you to zoom in on those data points that don't match your expectations, while ignoring everything that is already accounted for?

We call this attention. You pay attention to the unexpected bang, the unforeseen brush on your skin, the surprising movement in your periphery. Attending allows you to put your high-resolution sensors on the problem and figure out how to incorporate it into your model. *Ah, that's just the lawn mower, that's the kitten, that's a housefly.* Your model is now updated. In contrast, you pay no attention to the feeling of the shoe on your left foot, because you already have an internal model of it, and that model is consistently predicting what you're receiving. At least, until you get a pebble in your shoe. That draws your attention, because suddenly the model calls for an update.

The difference between predictions and outcomes is the key to understanding a strange property of learning: if you're predicting perfectly, your brain doesn't need to change further. Say you learn that a ding of your phone predicts that you just got a text message. Your brain will quickly learn the relationship between the two, in large part because of the relevance of text messages to your social life. Then let's say your phone gets a software update, and as a consequence the arrival of a text message is now signaled by a ding *plus* a vibration. It turns out that your brain won't train up on the vibration; this is an effect known as *blocking*. Your brain already knows that the ding predicts the text, so it has no need to learn about something new. If your phone merely vibrates, without the ding, your brain won't know the meaning of the cue; it has learned nothing about that.[10] The existence of blocking makes sense when we understand that changes in the brain happen only when there's a difference between what was expected and what actually happens.

Our internal model of the world allows us to make predictions and quickly detect when we're wrong—which tells us where to attend and how to update. And this sort of system is becoming interesting to engineers thinking about the future of machinery: several companies are starting to work on devices that operate this way, from tractors to airplanes. An internal model of the world allows a machine to make its best guesses about the events that are expected to unfold. When events are consistent with the predictions of the machine's algorithms, nothing has to change. It is only when the inputs go off script that the software needs to pay attention to update the machine's model.

With this background in hand, it is easy to understand how drugs modify nervous systems. Consumption of a drug changes the number of receptors for the drug in the brain—to such an extent that you can look at a brain after a person has died and determine his addictions by gauging his molecular changes. This is why people become desensitized (or tolerant) to a drug: the brain comes to predict the presence of the drug, and adapts its receptor expression so it can maintain a stable equilibrium when it receives the next hit. In a physical, literal way, the brain comes to expect the drug to be there: the biological details have calibrated themselves accordingly. Because the system now predicts a certain amount to be present, more is needed to achieve the original high.

This recalibration is the basis of the ugly symptoms of drug withdrawal. The more the brain is adapted to the drug, the harder the fall when the drug is taken away. Withdrawal symptoms vary by drug— from sweating to shakes to depression—but they all have in common a powerful absence of something that is anticipated.

This understanding of neural predictions also gives an understanding of heartbreak. People you love become part of you—not just metaphorically, but physically. You absorb people into your internal model of the world. Your brain refashions itself around the expectation of their presence. After the breakup with a lover, the death of a friend, or the loss of a parent, the sudden absence represents a major departure from homeostasis. As Kahlil Gibran put it in *The Prophet,* "And ever has it been that love knows not its own depth until the hour of separation."

In this way, your brain is like the negative image of everyone you've come in contact with. Your lovers, friends, and parents fill in their expected shapes. Just like feeling the waves after you've departed the boat, or craving the drug when it's absent, so your brain calls for the people in your life to be there. When someone moves away, rejects you, or dies, your brain struggles with its thwarted expectations. Slowly, through time, it has to readjust to a world without that person.

GOING TOWARD THE LIGHT. OR SUGAR. OR DATA.

Consider phototropism in plants: the act of capturing maximum light by adopting new positions. If you watch a plant growing in fast motion, you will see that it doesn't grow straight toward the light source; instead, it overshoots its trajectory by a little bit, then undershoots by a bit, and so on. Instead of a preplanned mission, it's a spastic dance with constant correction.

A similar strategy is found in the movement of bacteria. When they are searching for the center of a food source—say, a bit of sugar that has fallen on the kitchen counter—they make their way to the sugar by employing three elegantly simple rules:

1. Randomly select a direction and move in a straight line.
2. If things are getting better, keep going.
3. If things are getting worse, randomly change directions by tumbling.

In other words, the strategy is to lock down the approach when conditions are improving and dump it when it's not working. By this simple policy, a bacterium can quickly and efficiently work its way to the densest point of the food source.[11]

I propose there's a similar principle at work in the brain. Instead of working its way toward maximizing sunlight or food, it works toward maximizing information. I call this strategy infotropism. This hypothesis suggests that neural circuitry constantly shifts to maximize the amount of information it can extract from the environment.

Consider what we've seen in the previous chapters. We witnessed the way a brain comes to employ its sensory organs, whether they capture photons, electrical fields, or scent molecules. We saw the way a brain comes to steer a body, no matter if that body possesses fins, legs, or robotic arms. Whatever the case, the brain fine-tunes its circuitry to maximize the data it streams from the world. The fine-tuning is helped along by rewards, which cause broadcasts throughout the circuitry to announce that something worked. In this way, with a minimum of preprogramming, the system works out how to optimize its interaction with the world.

For example, we saw how the neural landscapes carved themselves into baby Hayato in Osaka and into baby William in Palo Alto, allowing them to distinguish different sounds. I discussed this as an example of reward-based modification, but now we can see this from a higher level as infotropism: the babies' brains adjusted to maximize the data that mattered around them.

On a longer timescale, we saw that when a person goes blind, other senses take over the visual cortex. In the next chapter, we'll learn how neurons pull this off, but for now note that we can interpret this take-over as infotropism: the brain maximizes its resources to interpret whatever data flows in.

And recall the illusion with the horizontal and vertical colored lines. Your visual system works to separate the dimensions of color and orientation because it's trying to maximize information from the world. Accordingly, it doesn't want to mix together these separable measures. Although the effect is typically viewed as simply a fun visual illusion, the work happens under the hood for a deeper reason: if something were causing a tinge to appear on lines (say, strange overhead lighting or something wrong with your optics), your brain would reorganize itself to take care of this, canceling out the relationship. By doing so, it would maximize your capacity to extract information about colors and orientations separately. By separating two dimensions that (statistically) should remain unyoked, your brain can best gather information from the world.

Here's an example of infotropism at the level of neurons: your retina (at the back of your eye) reads the world differently in the day and in the night. In the noonday brightness there are plenty of photons to capture, and so each photoreceptor takes care of its own tiny dot of the scene, yielding high resolution. At night it's a different story. There are fewer photons to be snagged, so now the important part is detecting that *something* was there, even if it's without high spatial resolution. So the photoreceptors do their work at night in a very different manner, changing the details of their internal molecular cascades and joining forces with one another. Under these conditions, it takes longer to register there was something out there, but collectively they can detect much lower levels of light.[12] This sophisticated strategy allows the retina to operate differently as the light levels go up or down. When it's bright out, the system achieves high spatial resolution; when it's dark, photoreceptors pool together to have a better chance to catch photons, resulting in vision that's more sensitive to dim light but blurrier in resolution. The system puts enormous work into shifting itself to a point where it can maximize information. Whether photons are plentiful or rare, the retina optimizes itself to capture data. In the day it captures the most detail so that it can spot the rabbit at a distance; in dim light it shifts to higher sensitivity to capture whatever's out there with lower detail, capturing the shadowy essence of the jaguar lurking in the gloom. Mother Nature figured out not only how to build an eye but also how to adjust its circuitry on the fly so it can operate differently in various contexts— all to make the best use of what's available. It's infotropic.

ADJUSTING TO EXPECT THE UNEXPECTED

Just as the plant seeks sunlight and the bacteria seeks sugar, the brain seeks information. It tries to constantly change its circuitry to maximize the data it can draw from the world. To that end, it builds an internal model of the outside, which equates to its predictions. If the world proceeds as expected, the brain saves energy. Remember the soccer players earlier in the book: the amateur has a great deal of brain activity, while the pro has little activity. This happens because the pro has burned his predictions of the world directly into his circuitry; the amateur is still scrambling to make a reasonable forecast.

Fundamentally, the brain is a prediction machine, and that is the driving engine behind its constant self-reconfiguration. By modeling the state of the world, the brain reshapes itself to have good expectations, and therefore to be maximally sensitive to the unexpected.

And now we're ready for the next question: Given everything we've seen in the book so far, how is this all implemented at the level of cells in the brain?

8

BALANCING ON THE EDGE OF CHANGE

Imagine you're a space alien who happens to visit Earth in October 1962. You've arrived just in time for the tense standoff of the Cuban missile crisis.

You would be forgiven for thinking that nothing much is happening. To your googly eyes, the United States is not doing anything, and neither are the Cubans and Soviets. Yawning with a green hand over your mouth, you'd likely conclude that you are looking at a political system that is unmotivated, lethargic, or fossilized.

It might not strike you that the only reason nothing is happening is because all the counterforces are balanced in perfect opposition to each other. All the springs are tightly wound, the missiles are mutually aimed, the armies are at the ready.

Although it's not always easy to see, this is the situation with the brain. It can appear that its maps are stable simply because they are perfectly balanced in their counterforces. The brain gives the illusion that it is settled into stillness, but the principles of competition poise it on the hair-trigger edge of change. Don't let the calm fool you: neural networks appear settled only because every region is trapped in a cold war, tightly wound, ready to compete for the future borders of the internal globe.

WHEN TERRITORY DISAPPEARS

The countries of Haiti and the Dominican Republic share the Caribbean island of Hispaniola. Consider what would happen if a tsunami were to slam into the Dominican Republic and make it uninhabitable. One possibility is that the Dominicans would be erased from the map and Haiti would continue business as usual. But there's a second possibility: What if the Haitians shifted their nation several hundred miles to the west, bigheartedly accommodating the Dominicans by shrinking their own territory and sharing what remained? In this case, thanks to neighboring generosity, the two nations would be harmoniously compressed onto a smaller, remaining bit of real estate.

Returning to the brain: What happens when disease, surgery, or brain damage result in less available territory? Just as with neighboring countries, there are two possibilities. The brain might leave out the parts of the map corresponding to the missing tissue, or the brain might squish the original map on a smaller piece of real estate.

To determine which it is, let's turn to a young girl we'll call Alice. At the age of three and a half, she began to have small seizures, so her parents took her to the hospital for a brain scan. To the surprise of the medical community, they discovered that she had been born with *only* the left half of her brain. In a rare abnormality, the right half was simply missing.[1]

But here's the surprise: she had a normal childhood. Amazingly, talents such as eye-hand coordination were unaffected by her strange developmental twist. She had seizures, but these were controllable by medications, and soon the only outward sign of Alice's missing hemisphere was her difficulty with fine motor movements using her left hand.

Alice's condition gives us a chance to ask a fundamental question: What happens to brain wiring that is normally distributed across two hemispheres when only one hemisphere develops?

To understand the answer, first consider how information is normally carried to the brain from a person's left eye. Fibers from the left half of the retina carry their information to the back of the left visual cortex. No problem here, because Alice has the left side of her brain. But information from the *right* half of the retina normally crosses the midline, connecting to the back of the *right* hemisphere. Because Alice was missing that half of the brain, where would the fibers go?

In a marvelous example of livewiring that would never have been suspected in previous decades, fibers from *both* visual fields plugged

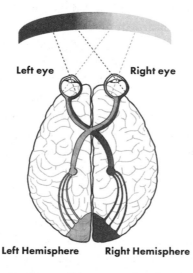

Normally the left side of the world is represented on the right side of the brain. Because Alice's right hemisphere was missing, where would the fibers go?

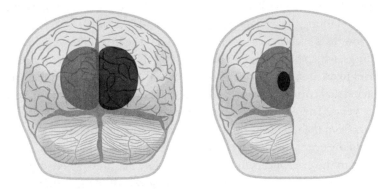

The visual cortex at the back of the brain. Left, *Typical brain. Gray shows where the right visual field (right side of the visual world) is represented, and black shows where the left visual field is represented.* Right, *Alice's visual system rewired to allow both the left and the right visual fields to be represented in the single remaining hemisphere.*

into the left hemisphere. The entire visual field became represented on the only available real estate. In other words, Haiti shared the territory that remained.

The fact that Alice has normal vision and normal eye-hand coordination tells us something else remarkable: even though the first stages of her visual system had not wired up with the typical organization, her surrounding brain areas had no trouble figuring out how to employ the unusual map. In other words, her visual cortex didn't need to play by the normal genetic playbook for the rest of the system to function. Consistent with what we've seen throughout this book, Alice's genetics didn't set up a fragile system that would fail with major deviations from the plan. Instead, her genetics unpacked a livewired system that would figure out how to get the job done.

While Alice was born missing a hemisphere, recall Matthew's story from earlier in the book: he had one hemisphere surgically removed. Matthew has a slight limp, but otherwise is able to live an independent life with no one the wiser. Just like Alice's remaining hemisphere, Matthew's was able to figure out how to perform the necessary tasks: the brain tissue rewired itself to maintain business as usual, even when the territory was radically changed. For both Alice and Matthew, the cerebral maps rewired themselves onto half the previous real estate while retaining their relationships, tasks, and functions.

How does such radical rewiring happen? The first hints were found in frogs, which have simpler visual systems. Nerves from the frog's eye travel to an area known as the optic tectum (roughly akin to the primary visual cortex in mammals)—the right eye to the left tectum, and vice versa. There the nerves insert themselves in an orderly fashion: fibers from the top of the eye connect to the top of the tectum, the left part of the eye to the left part of the tectum, and so on. Each fiber coming from the eye appears to have a preassigned address where it plugs into the target. So what happens if you remove half of the tectum during development, before the nerves arrive? The answer—analogous to Alice's brain—is that a full map of the visual field develops on the

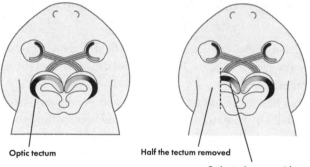

Optic tectum   Half the tectum removed

Retinotopic map squishes

Squishing to fit onto a smaller territory. Left, Normal mapping of the retina to the optic tectum. Right, With half the tectum gone, the map compresses to fit.

smaller target area.[2] The map looks normal. It's simply compressed, like the kindhearted map of Haiti after the eastern half of the island disappears.

Now comes the next level of the experiment: What would happen if you transplanted an *extra* eye onto one side of a tadpole? In this situation, an unexpected optic nerve now has to share the target of the tectum. What happens? The eyes divide up the territory in alternating stripes, each set of stripes containing a full map of the eye.[3] The incoming fibers once again utilize whatever space is available. It would

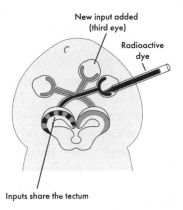

*If a third eye is transplanted, the tectum accommodates
the additional input in stripes.*

be as though a new country elbowed its way onto the island of Hispaniola, and Haiti voted to share the territory with the new kingdom
by compressing itself into alternating stripes.[4]

Such experiments demonstrate that maps can compress and share
territory when necessary. Can a map also *stretch* if more territory is
available? To find out, researchers removed one-half of the retina.
Now only half of the normal number of optic fibers reached the nor-

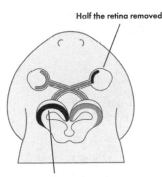

*With only half the retinal fibers arriving at the tectum,
the map stretches.*

mal sized territory of the optic tectum. What happens in this case? The map (now encoding only one-half of visual space) spreads out to utilize the entire tectum.[5]

The lesson from Alice, Matthew, and the frogs is that neural maps are not predefined by a genetic urban planning commission. Instead, whatever real estate is available gets used and filled.

This property of dynamic rewiring is the best hope after brain damage due to stroke. After swelling in the brain goes down, that's when the real work of the brain begins. Over the course of months or years, major cortical reorganization can occur, and functions that were lost can sometimes be regained. An example of this is often seen after a person loses the skill of language. In the majority of people, language is localized in the left hemisphere, and after a left-sided stroke they are no longer able to speak or understand words. However, language function will often begin to recover after some time—not because the dead tissue in the left hemisphere has healed, but because the job of language shifts over to the right hemisphere. In one report, two patients had left-hemisphere strokes followed by language impairment and eventual (partial) language recovery. But both of these unlucky patients then suffered right-sided strokes and showed a worsening of their recovered language, verifying that the function had transferred to the right hemisphere.[6]

So we see that brain maps stretch, squish, and relocate their functions. But how do they know how to do that? To answer that, we need to zoom in, deep into the forests of the neurons.

HOW TO SPREAD DRUG DEALERS EVENLY

I grew up in Albuquerque, New Mexico. The city has its share of physicians, lawyers, teachers, and engineers—and as everyone knows from the television show *Breaking Bad,* it also has its drug dealers. As I grew older, I began to wonder how each drug dealer found his own territory. After all, they're not distributed only in the poor neighborhoods (even though that's where most of the policing takes place); instead, they're sprinkled in every zone of the city, each controlling sales in a handful of blocks.

So how did they determine who operates which territory? There are two possibilities.

The first is that the Albuquerque city planners convened a council in which all the drug dealers were brought together, seated upon folding chairs in city hall, and distributed around the city in a fair and equitable manner. Let's call this the top-down approach.

The alternative approach is bottom-up. What if the dealers were competing with one another and the stakes were high? Through rivalry, each would figure out that there's only a certain amount of territory he could reliably control. With each doing his own thing, but fenced in by competition from his neighbors, the dealers would naturally find themselves distributed across the city.

What features would result from the bottom-up approach? Let's say a portion of Albuquerque gets destroyed by a tornado. What happens? After the city recovers emotionally, the drug dealers figure out how to compress their space, squeezing in a little tighter. No one has to give them directions: there is less territory available, and everyone has to divide it up.

In contrast, if Albuquerque's footprint suddenly doubles, we would find that the dealers spread out, filling the vacuum, taking advantage of more territory and reduced competition. No one has to tell them what to do.

The high-level patterns of the city emerge from the competition of individuals. Each dealer vies for business. Each has loved ones to support, rent to pay, and perhaps a car they want—so each chronically fights for his niche. The flexibility of the city's drug dealer map is an inadvertent consequence of individual behavior, not an ingenious design by the city planners.

Now let's return to the brain. Pick up any neuroscience textbook and you'll read about neurotransmission: the release of a small amount of a chemical messenger from a neuron. The chemical binds to receptors on another cell, causing a little blip of electrical or chemical activity. In this way, neurons send messages to one another.

But consider this cellular interaction in a different light. Throughout the microscopic kingdom around us, single-celled creatures release chemicals. But those chemicals aren't friendly messages; they're instead defense mechanisms, shots across the bow. Now think of the billions of cells in the brain as being billions of single-celled organisms. Although we typically think of neurons cooperating happily, we can also view them as being locked in chronic battle. Instead of transmitting information to one another, they're spitting on each other. Through this lens, what we witness in active brain tissue is the competition among billions of individual agents, each of whom is battling for resources and trying to stay alive. Like the Albuquerque drug dealers, each is self-interested.

In this light, certain experimental findings become straightforward to understand. For example, in the early 1960s, the neurobiologists David Hubel and Torsten Wiesel showed that alternating stripes in the visual cortex of mammals carry signals from either the left eye or the right eye. Under normal circumstances, each eye controls an equal split of the real estate. But if one eye is patched shut early in its life, the stronger input from the other eye begins to take over more territory. In other words, maps in the visual cortex can be drastically changed by experience: inputs from the strong eye are retained and strengthened while the inputs from the shut eye are weakened and eventually decay.[7] This demonstrated two things. First, that these maps are not purely innate. Second, maintaining territory in the brain is activity-dependent: preserving ground requires constant vigor. As inputs diminish, neurons change their connections until they find where the action is.

This insight (which won Hubel and Wiesel the Nobel Prize in 1981) tells us what to do with children with misaligned eyes. A child born wall-eyed or cross-eyed will eventually lose vision in the eye that is less used. But the problem is not in the eye itself; it's in the visual cortex. Because one eye is dominant, it outcompetes the misaligned one, taking over more territory at the back of the brain. The solution? Surgically fix the weak eye to make it align, and then cover up the child's good eye with a patch. This gives the weaker eye a chance to reannex its lost cortical territory.[8] Once the balance is restored, the patch is removed, and both eyes work equally well.

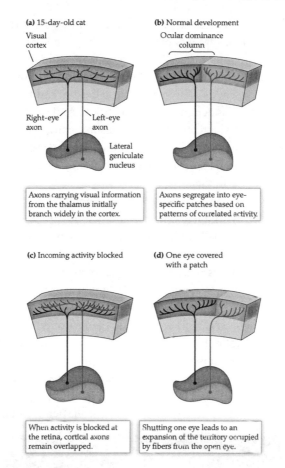

(a) 15-day-old cat

Visual cortex

Right-eye axon

Left-eye axon

Lateral geniculate nucleus

Axons carrying visual information from the thalamus initially branch widely in the cortex.

(b) Normal development

Ocular dominance column

Axons segregate into eye-specific patches based on patterns of correlated activity.

(c) Incoming activity blocked

When activity is blocked at the retina, cortical axons remain overlapped.

(d) One eye covered with a patch

Shutting one eye leads to an expansion of the territory occupied by fibers from the open eye.

Opposite page: (a) In a young animal, the input layer of primary visual cortex has uniform input from the left and right eyes. (b) As the animal matures, the connectivity from the two eyes comes to take up alternating regions. (c) If both eyes are deprived of light, the fibers carrying left- and right-eye information do not separate. (d) If only one eye is deprived of light, its inputs progressively shrink, while inputs from the working eye win more territory.

This useful trick falls out naturally from understanding the competition inherent at the level of neurons. And recall the brain's map of the body, the homunculus. The mystery we confronted in chapter 3 is how the brain (locked away in its dark cranium) knows what the body looks like. The lesson that surfaced from changes to the body plan is that the brain figures out the body map from simple rules. In other words, the map congeals naturally from interaction with the world, with adjacent areas of the body staking out adjoining representations in the brain.[9] Just like the drug dealers and the cross-eyed children, this process depends on competition. And that's why as soon as a limb goes missing (say, Admiral Nelson's arm), the neighboring cortical territory takes it over. Maintaining territory requires constant input to the individual neurons: when effort slows, they seek to switch teams to the active inputs.

This is also, by the way, why the homunculus looks like quite a strange person. The fingers and lips and genitals are huge, while the torso and legs are small. This results from the same sort of competition: there's a much higher receptor density in the fingers, lips, and genitals, and lower resolution in, say, the torso and thighs. The areas that send the most information win the largest representation.

So the right way to think about the system is that there is competition at small levels, and we get emergent properties (stretching, shrinking, sharing) at higher levels. As the local wars rage through a lifetime, the brain's maps are redrawn. And this is because each neuron confronts the same challenge as the urban drug dealer: find an open niche, and then spend all your time defending it. The constant fight for territory in the brain is live-or-die: each neuron spends its life fighting

for resources so it can survive. But what are they competing for? Cash is king for a drug dealer. What is the equivalent for a neuron?

———————

In 1941, a young Italian woman named Rita Levi-Montalcini fled from her native Turin into a small cottage in the country, where she lived in hiding from the Germans and Italians: her life was in danger because she was Jewish, and her home country had allied with the Nazis. While in hiding, she set up a small laboratory in the cottage and spent her days and nights trying to figure out how limbs developed in chick embryos. Her work there led to the discovery of nerve growth factor, and for this work she won the 1986 Nobel Prize.

What she had discovered was the first member of a class of life-preserving chemicals called neurotrophins.[10] These proteins, secreted by the neurons' targets, are the currency for which the neurons and synapses compete. It drives them to make and stabilize connections. Neurons that are successful at getting these life-preserving chemicals thrive. Unsuccessful neurons try reaching their branches elsewhere. If they are successful nowhere, they eventually die.

Beyond seeking the reward of these chemicals, neurons also avoid the danger of toxic factors. For example, synaptotoxins eliminate existing synapses,[11] and axons vie to escape these punishing effects by remaining active: as soon as they drop below a threshold, they are eliminated.[12]

In this way, the multilayered language of attractive and repulsive molecules provides the feedback that allows neurons to determine whether they should remain at their posts, blossom, shrink, slink somewhere else, or remove themselves for the sake of the common good.

———————

In parallel with the factors at the level of the individual neurons, a larger-scale issue determines whether the whole system is flexible or locked down. There are two types of neurons: those that transmit messages that stimulate their neighbors (excitatory) and those that thwart their neighbors (inhibitory). These two types of cells are interwoven in the networks, and together they determine how flexible the system is. If there's too much inhibition, neurons cannot adequately compete, and there is no more change. If there is too little inhibition, the competition is so high that a winner cannot emerge. A well-calibrated, flexible system requires just the right balance of inhibition and excitation; this way, neurons can square off with the proper amount of competition, a Goldilocks zone of not too little and not too much. As competition declines, the system solidifies into place. If competition rages too fiercely, winners cannot rise to the top.

As a metaphor, think of countries such as North Korea and Venezuela. North Korea has a regime in which the inhibition is so strict that people cannot do anything that isn't preapproved by the government. In Venezuela, the government has such a weak hold that drug cartels, mafias, and criminals run rampant. In both cases, the countries do not flourish—the first because of too much inhibition, the second because of too little. Around the world, productive nations keep themselves balanced at a sweet spot between being too malleable and being too rigid. Two-party systems are very useful for this, and for the present purposes think about conservatives and liberals as analogous to the two competing types of neurotransmission, inhibition and excitation. Typically, one of the parties dominates, but just barely. Often a president belongs to one party, while Congress follows the tune of the other. While it is common to lament the chronic debate that emerges from bipartisanship, the two-party system is ideal for making change when useful. In contrast, total domination by one party leads to a takeover of the system that historically spells woe for a nation.[13] The useful magic, in governments and in brains, comes from counterbalancing forces: that's how you keep a system composed, equilibriated, and ready to make change.

HOW NEURONS EXPAND THEIR SOCIAL NETWORK

We saw earlier that changes can happen rapidly in the brain—in as little as an hour. How do such large modifications occur so quickly?

Remember the blindfolded subjects whose visual cortex begins to respond to touch within an hour (chapter 3)? This time period is too rapid for new synapses to grow from touch and hearing areas into primary visual cortex, and this observation suggests that the connections were already there.[14] After all, many neural connections already exist, but are so inhibited that their presence has no function. A release from inhibition is the step that allows them to be heard.[15]

As an analogy, imagine a major disruption to your circle of friends. Due to a tragic misunderstanding at a party (where everyone was acting just as crazy as you were), you lose all your closest friends. Suddenly your social input is less than it used to be, and now you begin listening for signals from more distant acquaintances—people who have never before had a chance to command your full attention. Their voices were squelched by the strong relationships you had with your tightest friends. Now that these peripheral friends begin to be heard, you start to fill out your social life by cultivating and strengthening these weak connections.

As you might be able to guess from this analogy, the mechanism for unmasking is the release of inhibition that the strong connections had previously provided. In neural terms, the previous connections provided lateral inhibition, meaning that they squelched the activity of their nearest neighbors.[16] When the original inputs quiet down (even from a very short-term change, such as anesthetizing your arm or putting on a blindfold), fast changes result. Sometimes this is from changes in the cortex; other times it's from the disinhibition of already-existing neighboring connections from thalamus to cortex.[17] In other words, as a result of disinhibition, the widespread and previously silent projections become functionally operative.

Unmasking these connections is possible only because the brain is highly cross wired with redundant connectivity. This redundancy starts off strong and diminishes through time. For example, play a loud beep to someone, and use electrodes on the scalp (electroencephalography) to measure the brain's response. In a normal adult, the beep elicits an electrical response that can be measured clearly over the auditory cortex, but is smaller or absent over the visual cortex. Now compare this with what you would see in a six-month-old child: the response over the auditory and visual regions is close to the same. Why? Because the redundancy of connections in the infant's brain means the auditory and visual areas are not so different from each other.[18] Between the ages of six months and three years, there is a gradual decrease in the size of the measurable response over visual areas to a beep. The brain starts off heavily interconnected and prunes the overlap with time. However, that early cross wiring does not go away completely. Even in the adult brain, primary auditory fibers reach directly into primary visual cortex, and vice versa.[19] And this cross-stitching is what can allow rapid redeployment when necessary.

Unmasking silent connections isn't the only way that changes happen. At a slower timescale, the brain employs a different trick: the growth of axons into new areas followed by a blossoming of connections.[20] To continue the analogy about your friends, imagine that you begin to exchange an increasing number of messages with those acquaintances you never paid much attention to before. With time, given the unexpected room in your social calendar, these distant friends begin inviting you to dinners at their homes, and you become open to new friendships that you lacked room for before. You seek out and establish new connections that stem from more distant social circles. And so it goes with the brain: with enough time, areas cut off from communication sprout new connections.[21]

To summarize where we are so far: a general principle of reorganization is that the brain conceals a great many silent connections. These are normally inhibited and don't contribute much of anything. But they're available if needed in the future. Leveraging these, the brain can respond rapidly to changes in input. However, these silent connections are limited in number, and for longer, more widespread change a different approach is used: if short-term changes are found to be useful to the animal, then long-term changes (such as the sprouting of new synapses and growth of new axons) will eventually follow. Beyond these approaches, there's one more thing that helps the system shape itself: death.

THE BENEFITS OF A GOOD DEATH

When we think of Michelangelo crafting one of his statues, it's easy to imagine that he built up the marble masterwork bit by bit—fashioning each finger, the nose, the forehead, the flowing robes. But remember that he started with a giant block of marble. His creations emerged by taking stone away, not by adding anything. His masterworks pivoted on discovering what was already possible inside the block.

This is the same principle the brain uses on a longer timescale. After all, neurons live their lives in a perpetual state of looking for their right place. They put out feelers. If they're getting a good response, they keep it going. If they get the cold shoulder, they try their luck nearby with other neurons. At some point, if they're getting no positive feedback, they get the message that they simply don't belong.

There are two ways that cells die. If they don't get enough nutrients (say, blockage of an artery leaves the tissue starving for blood), then the cells die a sloppy sort of death, in which inflammatory chemicals leak out and cause damage in the local neighborhood. This is known as necrosis. But the second way cells can die is by apoptosis, in which they neatly commit suicide. They purposefully fold up shop, take care of their affairs, and consume themselves. Apoptotic cell death is not a bad thing. In fact, it is the engine for sculpting a nervous system. In embryonic development, the trajectory from a webbed hand to clearly defined fingers depends on sculpting away cells, not adding them. The same principles apply to sculpting the brain. During development, 50 percent more neurons than needed are produced. Massive die-off is standard operating procedure.

IS CANCER AN EXPRESSION OF PLASTICITY GONE AWRY?

I think it's possible that our society's study of cancer will end up overlapping with our study of plasticity.

Here's the cartoon version of cancer: A cell obtains a mutation that causes it to divide over and over. With its replication out of control, it becomes a tumor and compromises the rest of the system.

But real cancer is more complex than that. In a tumor, billions of cells compete for survival, and tumor cells can be quite different from one another. Just as in the brain, these cells are locked in competition for survival. There are limited amounts of nutrients, and each cell is fighting for survival. Typical cancers involve the mutation of a cell that gives it a slight advantage in this do-or-die competitive crucible.[22] The advantage can be slight, something that allows it to slightly outcompete its nearest neighbors. However, once this new mutant cell has replicated, its own cells are now fighting with themselves. So further mutations can occur, giving a new advantage, making the new progeny just slightly better competitors. These continue to fight, evolve, and get to be better fighters, and eventually the tumor kills the host.

Now let's return to the brain and body. We are livewired creatures. Neurons in the brain (and more generally, all cells in the body) are locked in a battle for survival, and sometimes that fever pitch of competition will tip into pathology. Some mutations can give a slight advantage to a cell in this environment—but at the cost of tipping the whole system into a death spiral.

I suggest that multicellular organisms find their evolutionary niche on the knife-edge of chaos, trying to balance between competition that yields something useful and competition so fierce it kills the system. To my mind, this is one way to understand the enormous incidence of cancers in animals. Most mammals, for instance, have about a 30 percent chance of toppling into cancer by the end of their lives. It seems like a surprisingly easy state for the system to stumble into.

In a system that's wound so tightly for competition, a tiny advantage can turn disastrous. It can ratchet up the competition for mutations. A system in which all the pieces get along would not, presumably, lead to cancers with multiple mutations; it wouldn't need to.

SAVING THE BRAIN FOREST

We saw in this chapter how the simple rules of competition for territory allow the brain to encode maps that can be stretched and squished. We met Alice, who was born without one hemisphere, and we recalled Matthew, the boy who had one hemisphere removed. They both had rewired brains, such that both visual fields carried information to a single remaining hemisphere. This was made possible by competition at the level of the synapses and neurons, which allowed the rapid unmasking of existing connections as well as, with time, the growth of new axons and sprouting of new synapses. Throughout, Alice's and Matthew's desire to walk, to play tag, and to ride a bicycle provided the signals of relevance that allowed their brains to reorganize.

The complexity of what's found in a rain forest makes me contemplate the complexity of what's found in the brain forest. We tend to think of our eighty-six billion neurons like trees and bushes that all get along. But what if our neurons are really like the members of a forest, which are in constant competition to stay alive? Trees and bushes try endless strategies to grow taller, or wider, or otherwise outcompete one another, because they're all trying to get exposure to sunlight. Without light, they die. The neurotrophic factors we saw are the sunlight to the neurons, and someday we may understand neurons' strategies in the terms of the competitive tricks they play on one another.

As highlighted before, everything we saw in this chapter is fundamentally different from the way we build our current technologies. Engineers brag about intuitions of efficiency, minimum requirements, and cleanliness. Such a devotion to trimness achieves less wiring. But it also builds in an inability to balance on the edge of chaos, to be primed for the unexpected, to implement rapid change in the system.

With these elements in place, we're now ready to turn to a question that's been lurking in the background: Why are young brains so much more plastic than adult brains?

9

WHY IS IT HARDER
TO TEACH OLD DOGS NEW TRICKS?

BORN AS MANY

In the 1970s, the psychologist Hans-Lukas Teuber of MIT got curious about what had happened to soldiers who had sustained head injuries in World War II, almost thirty years earlier. He tracked down 520 men who had incurred brain damage during the battles. Some had fared well in their recovery; others showed poor outcomes. Scouring his paperwork, Teuber figured out the variable that mattered: the younger the soldier was when he got injured, the better he was now; the older the soldier, the more permanent the damage.[1]

Young brains are like a globe of the earth five thousand years ago: different events have the capacity to push the borders in many different directions. But today, after millennia of history, the maps of the globe are more settled into place. Now that humans have had centuries to clang swords and discharge rifles, territory borders are stubborn to change. Roving bands of marauders and horse-straddling conquerors have been replaced with the United Nations and international rules of engagement. Economies have grown increasingly dependent on information and expertise rather than on pillagable treasures. Moreover, nuclear arms have made it more forbidding to start a skirmish. Thus, even in the face of trade arguments and immigration debates, the boundaries between countries are unlikely to move. Nations have settled into place. The landmasses of our planet began with enormous possibility for the locations of borders, but with time the potential has narrowed.

The brain matures like the planet. Through years of border disputes, neural maps become increasingly solidified. As a result, brain injury is terribly dangerous for the elderly, while it is less dangerous for the young. An older brain cannot easily reassign settled territories for new tasks, while a brain at the dawn of its wars can still reimagine its maps.

Let's return to baby Hayato and baby William. When they are first born, they can understand all the sounds of human languages. And they can also do much more: they pick up on the subtle details of their cultures, absorb religious beliefs, and learn the rules of social interactions. They learn how to gather massive amounts of information, and depending on their generation, they do so by unrolling a scroll, flipping pages of a book, or swiping the screen of a small rectangle.

But by the time they are grown, the story has changed somewhat. Hayato belongs to a particular political party and is unlikely to change. William plays the piano reasonably well, but has no particular interest in studying violin or other instruments. Hayato likes to cook, and all of his dishes exploit combinations of the fourteen ingredients he is used to. William spends his online time with a vanishingly small fraction of the billions of available websites. Hayato commands a respectable golf game, but holds no curiosity about other sports. William lives in a city of eight million people, but has only three close friends. Hayato is not particularly interested in the science he didn't already learn in school. In the store, William passes racks of shirts until he finds the kind he always wears, and he selects two in his standard colors. Hayato's haircut remains the same as it has been since he was eight years old.

These life trajectories underscore a general point: human babies are born with few built-in skills and a great deal of plasticity, while adults have mastered specific tasks at the expense of flexibility. There's a trade-off between adaptability and efficiency: as your brain gets good at certain jobs, it becomes less able to tackle others.

Recall the story in chapter 6 about the violinist Itzhak Perlman, in which a fan told him he would give his life to play like that, and Perlman replied "I did". Perlman was pointing to a fact of life: to get good at one thing is to close the door on others. Because you possess only a single life, what you devote yourself to sends you down particular roads, while the other paths will forever remain untrodden by you. Thus, I began this book with one of my favorite quotations from the philosopher Martin Heidegger: "Every man is born as many men and dies as a single one."

From the point of view of your neural networks, what does it mean to descend into pattern and habit? Imagine two towns a few miles apart. People interested in caravanning from one settlement to the other take all possible paths: some travelers walk the scenic route along the ridgetops, some prefer the shade of the cliffside, some move among slippery rocks by the river, and others take the riskier but faster route through the woods. With time and experience, one route proves more popular. Eventually the path becomes grooved where the most people have walked, and it starts to become the standard. After some years the local government lays down roadways. After some decades, this expands into highways. Broad optionality reduces to the standard.

Similarly, brains begin with many possible routes through the neural networks; with time, the practiced pathways become difficult to exit. Unused paths become thinned away. Neurons that can't find success with the world eventually fold up shop and commit suicide. Through decades of experience, the brain comes to physically represent the environment, and your decisions follow the remaining, hard-paved paths. The upside is that you end up with lightning-fast ways of solving problems. The downside is that it's harder to attack problems with wild unstructured inventiveness.

Beyond the diminishing optionality in the pathways, there is a second reason that older brains are less flexible: when they change, they do so only in small spots. In contrast, baby brains modify across vast territories. Using broadcast systems like acetylcholine, infants transmit announcements throughout the brain, allowing pathways and connections to modify. Their brains are changeable everywhere, slowly coming into focus like a Polaroid photograph. An adult brain changes only little bits at a time. It keeps most of its connections locked into place, to hold on to what has been learned, and only small areas are made flexible via a combination lock of the right neurotransmitters.[2]

An adult brain is like a pointillist artist who modifies the hue of only a few dots in an almost-finished painting.

As a side note, you may wonder what it *feels* like to be inside the massively malleable brain of a baby. We've all been there as infants, but we can't remember it. So what is it like to be plastic, uninhibited, and learning about a wide range of novel events? You can probably get close to understanding it by considering other situations in which your awareness and plasticity are firing on all cylinders. When you're an alert traveler in a new land, you drink in the sights of the foreign country, experiencing more novelty, more learning, and more distributed attention. After all, at home you pay attention to very little; it is all so predictable. When you are the traveler, you overflow with consciousness. In this view, when we are highly engaged and paying attention, we are like babies again.[3]

The differences between a baby and an adult are easy to see, but the neural transition from one to the other does not happen in a smooth line. Instead, it is like a door that swings closed. Once it shuts, large-scale change is over.

THE SENSITIVE PERIOD

Consider Matthew, the child we met at the beginning of the book who had half of his brain surgically removed. This kind of radical procedure, known as a hemispherectomy, is generally recommended only if the patient is less than eight years old. Matthew was six when he went under the knife—nearing old age for this surgery. If a child is older (say, an adolescent), he will have to function in life by bending tasks to fit his brain, rather than counting on his brain to adapt to the tasks.[4]

The closing door can be seen with more subtle changes to the brain. Recall Danielle, the profoundly neglected girl discovered in a Florida home. Locked in a small room throughout her childhood—without conversation and affection—she ended up incapable of speech, long-distance sight, and normal human interaction. Danielle's prospects for recovery are poor, and that's for one main reason: she was discovered too late. By the time the police found her, her world map had largely stabilized into place.

Matthew and Danielle divulge the same story: brains are most flexible at the beginning, in a window of time known as the sensitive period.[5] As this period passes, the neural geography becomes more difficult to change.

As seen with Danielle, a child's brain needs to hear plentiful language during a sensitive period. Without such input, the neurons never arrange themselves to capture the fundamental concepts of language. You may wonder what happens with a deaf baby, who doesn't hear auditory input. The answer: as long as the parents present sign language to the baby, her brain will wire up correctly for communication. Additionally, the deaf baby will employ her hands to babble, making resemblances to sign language—in the same way that a hearing baby exposed to language will babble with her vocal cords.[6] If there's input to pick up on, the baby will do so, as long as that input arrives within the sensitive period. After that door swings shut, it becomes too late to learn the fundamentals of communication.

So there's a window for acquiring the ability to communicate, and also for more subtle aspects of language, such as accents.[7] The actress Mila Kunis speaks American English with no discernible accent, so most people don't know she was born in Ukraine and lived there, not speaking a word of English, until the age of seven. In contrast, Arnold Schwarzenegger, who has been in contact with Hollywood and American filmmaking since his early twenties, has little hope of shaking his Austrian accent. His use of English began too late, brain-wise. Generally, if you arrive in a new country during your first seven years, your fluency in the new tongue will be as high as a native speaker's, because your window of sensitivity for obtaining the sounds is still open. If you immigrate when you're eight to ten years old, you'll have a slightly more difficult time blending in, but you'll be close. If you're past your teen years when you move, like Arnold was, your fluency is likely to remain low, and you'll be stuck with an accent that reveals your history. Your ability to sonically morph into a different culture is a door that remains open for only about a decade.

We saw earlier how we could leverage the principles of competition to help a child born with misaligned eyes: cover the good eye for a while, allowing the weak eye to fight to regain its lost territory. But note that the good eye has to be patched within the sensitive period—about the first six years—otherwise it's too late: the vision will never again be recoverable.[8] After six years, the dirt roads in the brain have been paved into highways and cannot now be modified.

The same lessons apply to blindness. As we saw earlier, the amount of takeover of the visual cortex is greatest for those born blind, less for those who go blind in early childhood, and the least for those who go blind later in life. Earlier changes in input are more easily dealt with than later changes. This principle can also be seen through the lens of performance. The more takeover of the visual cortex, the better a person can remember lists of words, because the former visual cortex is now being leveraged, in part, for memory tasks.[9] As expected, the memorizing benefit is highest in those who are born blind. The second-best performers are the early-blind, and the memorization improvements are smaller or absent in the late blind.[10] Timing matters.

This kind of knowledge is critical to surgeons when weighing possible procedures. A surgery to correct a blockage of the eyes can have very different outcomes depending on the age of the patient: a young person can redevelop visual experience quickly; not typically so for an older person. In fact, for people who have been blind for a long time, re-plugging visual data into the occipital cortex can sometimes disrupt a stable system for touch and hearing.[11]

Let's return to the experiment in which the visual nerves of a ferret were rewired to plug into its auditory cortex. Although the visual information was coming into an unusual region, the cortex figured out how to analyze the data. However, note that the transformation of the auditory cortex was not complete: the wiring ended up slightly more disorganized than it does in the visual cortex, raising the possibility that the auditory cortex is inherently optimized for slightly different input.[12] This may mean that the cortex's capacity for change is counterbalanced by at least some genetic pre-specification. But it may equally mean that by the time the experimental manipulation took place, the auditory cortex had already had a chance to lay down some statistics of the sounds around it. If visual fibers had been connected up from the very first moments of development (say, in the womb—a currently impossible experiment), it's possible that the transformation would be complete.

The influence of developmental timing is found across all the senses. Remember how the body maps readjust when a finger goes missing or when a new instrument is learned? Across the board, such change happens more in young brains than in old brains. Just like Mila Kunis and her unaccented speech, so we find that Itzhak Perlman took up the violin at a young age. If you were to take up the violin for the first time in your teens, there is no possibility that you would become a Perlman. Even if you worked harder to rack up the same number of hours of practice, your brain is already behind in the race: it has grown too solidified by the time of your first adolescent pizzicato.

Acquiring vision and language and violin proficiency depends on normal input from the world, and if a child such as Danielle doesn't receive these, she cannot later. The ability to learn language, possess vision, interact socially, walk normally, and have normal neurodevelopment is limited to the years of young childhood. After a certain point, these abilities are lost. The brain needs to experience the proper input within the right window to achieve its most useful connectivity.

As a result of the diminishing flexibility, we are highly influenced by events in our childhoods. As an interesting example, consider the correlation between how tall a man is and how much salary he will command. In America, each additional inch of height translates into a 1.8 percent increase in take-home pay. Why is this? The popular assumption is that this stems from discrimination in hiring practices: everyone wants to hire the tall guy because of his commanding presence. But it turns out there's a deeper reason. The best indicator of a male's future salary is how tall he was *at the age of sixteen*. However tall he grows after that doesn't change the outcome.[13] How do we understand this? Could it be some effect of nutritional differences between people? No: when the researchers correlated with height at ages seven and eleven, the effect was not as strong. Instead, the teenage years are a time when social status is being worked out, and as a result who you are as an adult strongly depends on who you were then. In fact, studies that track thousands of children into adulthood find that socially oriented careers, such as sales or managing other people, show the strongest effect of teenage height. Other careers, such as blue-collar work or artistic trades, are less influenced. How people treat you during your formative years has a great deal of impact on your comportment in the world, in terms of self-esteem, confidence, and leadership.

As an example, the media star Oprah Winfrey is worth $2.7 billion—so it seems a little surprising that she is reported to have a deep-rooted fear of ending up homeless and penniless. But it's because of the path that got her here. Before she was an empress of the media, she was an impoverished child in Mississippi, born to a teenage single mother.

As Aristotle noted twenty-four hundred years ago, "The habits we form from childhood make no small difference, but rather they make all the difference."

To capture the idea of the sensitive period, I introduced the metaphor of a door swinging shut. But now we're ready to take the analogy to the next level. It's not one door: it's many.

DOORS CLOSE AT DIFFERENT RATES

The brain is so impressionable in its earliest days that it can sometimes get into hot water. For example, the baby goose hatches from its egg and establishes a parental relationship with the first animate object it sees. This is a sufficient strategy in most cases—because that first sight is usually its mother—but it can get fooled in the wrong circumstances. In the 1930s, the zoologist Konrad Lorenz did not have to work hard for geese to imprint on him; instead, he just had to show up during a small window of plasticity after they hatched, and then they would imprint on him and follow him around.

It's a fast-swinging door for geese to imprint on their parent. But the geese can still learn other things later in life, such as where the river

Konrad Lorenz and his impressionable geese.

is, where to best seek food, and the identities of other geese they meet in adulthood.

Sensitive periods are different for different tasks of the brain. Not all brain regions are equally plastic in terms of how flexibly they begin and how long they retain their adaptability.

Is there a pattern to which areas solidify first? Consider what happened when researchers searched for changes in the adult visual cortex after damage to the retina. Would neighboring regions of visual cortex take over the unused tissue, and if so, how rapidly? To their surprise, there were no measurable changes in visual cortex. The part of the cortex that was inactive stayed inactive: it was not taken over by surrounding areas.[14] Given the history of plasticity studies, the answer was a little unexpected. After all, there's a good deal of flexibility in the somatosensory and motor areas of adults, allowing you to learn how to hang glide or snowboard even in your later years.[15]

So what was the difference between the studies involving your vision versus your body? Why are the patterns in the primary visual cortex locked into place after a short window of a few years, while the somatosensory and motor cortices can continue to learn? Why does an eight-year-old whose eyes are misaligned become irrecoverably blind in one eye, while a fifty-eight-year-old with paralysis can learn to control a robotic arm?

Different areas of the brain operate on different schedules of plasticity. Some neural networks are unyielding, while others are highly pliable; some sensitive periods are brief, while others are long.

Is there a general principle at work behind this diversity? One possibility is that the different sensitive periods are caused by different underlying learning strategies of different regions.[16] In this view, some regions are geared to learn throughout life, because they are meant to encode changeable details of the world. Think of vocabulary words, the ability to learn new directions, or the visual recognition of people's faces: these are tasks for which one wants to retain flexibility. In contrast, other brain areas are involved in stable relationships—such as the building blocks of vision, how to chew food, or the general rules of grammar—and these areas require a faster lockdown.

But how could the brain know in advance the order in which to solidify things? Is it genetically encoded? Possibly some aspects are, but I suggest a new hypothesis: the degree of plasticity in a brain region reflects how much its data change (or are likely to change) in the outside world. If the incoming data are unwavering, the system hardens around them. If the data are constantly changing, the system remains flexible. As a result, stable data solidify first.

Take information from the ears versus information from the body. Areas encoding the basic sounds of the world—such as the primary auditory cortex—become resistant to change. They stiffen rapidly. This is what happened to baby William and baby Hayato when they fixed into place their landscape of possible sounds. In contrast, motor and somatosensory areas involved in navigating the body remain more plastic, because body plans change throughout a lifetime: you get fat, thin, put on boots or slippers or crutches, hop on a bike or a scooter or a trampoline. That's why adult William and adult Hayato can meet for a vacation in which they successfully learn windsurfing. While the statistics of sound don't change much, your body's feedback from the world constantly does. As a result, the primary auditory cortex tightens down; less so for the body plan.

Let's zoom in to a single sense, like vision. In low-level visual areas—such as the primary visual cortex—neurons encode basic properties of the world: edges, colors, and angles. In contrast, higher areas of visual cortex are involved with more particular items, such as the layout of your street, or the sleek look of this year's sports car, or the arrangement of apps on your phone. The information in the low-level areas becomes established first, and successive layers wire up on top of those foundations. So the possible angles at which lines can be oriented are fixed into place, but you can still learn the face of the latest movie star. In this hierarchy of flexibility, the representations at the bottom are learned first; these reflect the basic statistics of the visual world, which are unlikely to change. These lower-level representations remain stable so that higher-order assemblies (which may change more rapidly) can be learned.

By analogy, if you're building a library, you'll want to nail down the basics first—establishing the positions of the shelves, the Dewey decimal system for organization, and the work flow for checking out books. Once those are nailed down, it is straightforward to maintain a flexible inventory of books, expanding the offerings in exciting categories, reducing outdated volumes, and constantly testing new titles.

So there's no single answer to whether the brain is plastic as we get older. It depends on what brain area we're talking about. Plasticity declines with age, but across the brain it declines differently, steeply or shallowly, depending on its function.

This plasticity-reflects-variance hypothesis has an analogy in genetics. In ways that science is still working to understand, genomes seem to lock in some parts of their nucleotide sequences more than others, protecting them against mutation. Conversely, other regions of the chromosomes are more variable. Roughly speaking, the variability of a genetic sequence mirrors the variability of features in the world.[17] For example, skin pigment genes are variable, because humans find themselves at different latitudes and need to change pigmentation to absorb enough vitamin D. In contrast, genes coding for proteins that break down sugar are stable, because that is a critical and unchanging energy source. By analogy, future research may be able to quantify the "variability" of mental, social, and behavioral functions in human life, and put to the test the hypothesis that the most flexible circuits of the brain mirror the most variable parts of our environment.

STILL CHANGING AFTER ALL THESE YEARS

Adults envy children. Children have the ability to absorb languages at an extraordinary rate, to think of magically bizarre approaches to a problem, and to celebrate the novelty of every experience—from looking out an airplane window to petting a rabbit for the first time. Older brains have more closed doors, which is why Teuber's World War II veterans fared worse if they were older and why Schwarzenegger retains his thick accent. Similarly, the older a city is, the more its infrastructure becomes resistant to shift. Rome, for example, cannot untangle its windy roads to resemble the gridwork of Manhattan; too much history has glued its snaking routes into place. Just like developing humans, cities deepen their tracks along early roads.

In 1984, at the age of thirty-five, the physicist Alan Lightman wrote a short essay in *The New York Times* titled "Elapsed Expectations" in which he lamented the perceived stiffening of his mind:

> The limber years for scientists, as for athletes, generally come at a young age. Isaac Newton was in his early 20's when he discovered the law of gravity, Albert Einstein was 26 when he formulated special relativity, and James Clerk Maxwell had polished off electromagnetic theory and retired to the country by 35. When I hit 35 myself, some months ago, I went through the unpleasant but irresistible exercise of summing up my career in physics. By this age, or another few years, the most creative achievements are finished and visible. You've either got the stuff and used it or you haven't.

These same sentiments were echoed by the physicist James Gates in a television interview:

There's a saying that old physicists accept new ideas when they die. It's the next generation that brings new ideas to their full fruition. When you get to be an old physicist like me, you know a lot of stuff, and it acts like a ballast on a ship; it pulls you down. You have all the weight of these other things that you know. And sometimes an idea, like a small fairy or a sprite, passes by and you say, "Ah, I don't know what that is, but it can't be very important." Well, sometimes it is.

Such dirges are typical of the aging. But happily, although brain plasticity diminishes over the years, it is still present. Livewiring is not solely the privilege of the young. Neural reconfiguration is an ongoing process that lasts through our lives: we form new ideas, accumulate fresh information, and remember people and events. Despite decreased flexibility, Rome evolves. The city now isn't what it was twenty years ago. Today its statuary is ringed with cell phone towers and internet cafés. Although the rudiments are difficult to change, the city nonetheless advances its finer points according to new circumstance—just as the library changes its stock while its architecture remains largely set.

We've seen this with many studies throughout the book—for example, with juggling, new musical instruments, maps of London, and so on—all of which involve adult plasticity. A stunning example emerged recently from the Nun Study, a multi-decade investigation of hundreds of Catholic nuns living in convents.[18] All the sisters agreed to regularly test their cognitive function, share their medical records, and donate their brains after death. Amazingly, some of the nuns never displayed any cognitive decline—they were sharp as a whip—and yet their brains at autopsy were riddled with the ravages of Alzheimer's disease. In other words, their neural networks were physically degenerating, but their performance was not. What could explain this? The key is that the nuns in their convents had to consistently use their wits until their final days. They had responsibilities, chores, social lives, arguments, game nights, group discussions, and so on. Unlike typical octogenarians, they didn't have a retirement that plopped them onto a couch in front of a television set. Because of their active mental lives, their brains were forced to constantly build new bridges, even as some of their neural roadways were physically falling apart. In fact, a stunning one-third of the nuns seem to have had the molecular pathology of Alzheimer's without the expected cognitive symptoms. An active mental life, even in the very elderly, fosters new connections.[19]

So learning can happen at any age. But why is it slower as a brain matures? One reason is that many of the swinging doors have closed. But there's another way to look at this. Remember that brain changes are driven by the *difference* between the internal model and what happens in the world. Thus, brains shift only when something is unpredicted. As you age and figure out the rules of the world—from the expectations of your home life to behavior in your social circles to the foods you prefer—your brain becomes less challenged with novel stimulation, and therefore more settled into place. For example, when you're a child, your internal model is that all people believe everything you believe. As world experience teaches you the difference between your prediction and your experience, your networks adjust to address the growing gap.

Or consider what happens when you start a new job. At first, everything is new—from your co-workers to your responsibilities to your approaches. You have a good deal of brain plasticity during the first days and weeks as you incorporate your new gig into your internal model. After a while, you become proficient at your job. Skill replaces flexibility.

We see this pattern in the way nations settle into place. Consider the amendments to the constitution of any country: almost all the change happens near the beginning, while the nation is learning the strategies of running itself; later, constitutions congeal into place and amendments slow. In the U.S. Constitution, for example, twelve of the amendments took place in the first thirteen years. After that, there were a maximum of four changes in any twenty-year period, and most periods had no changes at all. The latest change, ratifying the Twenty-Seventh Amendment, was in 1992. The Constitution has been at a standstill since then. In this way, nations steadily diminish their adaptation to the world: they profusely modify at the beginning, and with time they settle upon a working model that offers what the country needs to be operational.

In this same way, the brain's solidification reflects its success in understanding the world. Neural networks lock themselves more deeply into place not because of fading function, but instead because they have had success in figuring things out. So would you really want the plasticity of a child again? Although having a sponge-like brain that absorbs everything sounds appealing, the game of life is largely about figuring out the rules. What we lose in modifiability we gain in expertise. Our hard-won networks of association may not be fully correct or even internally consistent, but they add up to life experience, know-how, and an approach to the world. A child simply does not have the capacity to run a company, enjoy deep ideas, or lead a nation. If plasticity didn't decline, you would not lock down the conventions of the world. You'd never get good pattern recognition or the capacity to navigate a social life. You wouldn't be able to read a book, hold a meaningful conversation, ride a bike, or get food for yourself. Preserving total flexibility would retain the helplessness of an infant.

And what about the memories of your life? Imagine you could swallow a capsule that would renew your brain plasticity: this would give you the capacity to reprogram your neural networks to learn new languages rapidly and adopt new accents and new views of physics. The cost is that you'd forget what came before. Your memories of your childhood would be erased and overwritten. Your first lover, your first trip to Disneyland, your interaction with your parents—all would fade like a dream after waking. Would it be worth it to you?

A horror scenario about the future of warfare is to imagine a biological weapon that implements plasticity again: no one is physically hurt, but the troops are propelled back to the state of infants. They forget their ability to walk and to talk. All of their memories are wiped. When they are returned home by their commanders, they hold no remembrance of their families, friends, spouses, or children. Technically they are fine: they can still learn again; nothing is damaged. Only their mental lives—the part we cannot easily see—have had a factory reset back to their original state.

This scene is so horrific because, fundamentally, *who you are* is the sum total of your memory. We turn to that now.

10

REMEMBER WHEN

That last day the agony was perpetual. Time after time it lifted her almost off the bed, so they had to fight to hold her down. He could not endure and left the room; wept as if there never would be tears enough.

Jeannie came to comfort him. In her light voice she said: Grandaddy, Grandaddy don't cry. She is not there, she promised me. On the last day, she said she would go back to when she first heard music, a little girl on the road of the village where she was born. She promised me. It is a wedding and they dance, while the flutes so joyous and vibrant tremble in the air. Leave her there, Grandaddy, it is all right. She promised me. Come back, come back and help her poor body to die.

—TILLIE OLSEN, "TELL ME A RIDDLE"

Tillie Olsen's portrait of a dying grandmother tells of a woman whose recent memories have disappeared, even while her childhood memories remain rich and available. If you've known someone with dementia, you'll have noticed this pattern.

It's one of the oldest patterns noted in neurology. In 1882 this observation was canonized by the French psychologist Théodule Ribot, who was struck by his observation that older memories are more stable than newer memories.[1] This is known today as Ribot's law, and it explains why some people, as they reach the end of their lives, revert to their childhood language. In 1955, when Albert Einstein died in a hospital in Princeton, New Jersey, he spoke his final thoughts. Everyone wanted to know what the great physicist's last words were, but we will never know. It's not because there wasn't a nurse present to hear the final words, but instead because the words were spoken in German, his native tongue. The night nurse spoke only English, so his final utterances were lost.

No wonder Ribot was struck by the strangeness of this pattern of memory: other storage systems don't work that way. Institutional memory forgets old eras of leadership, educational institutions focus on recent trends, city governments brag of their latest accomplishments more than they dwell on successes from the past century.

So why does the brain do it backward? Why do older memories grow more secure? It's a critical clue to understand the principles running under the hood. And so we now turn to one of the most important aspects of livewiring: the phenomenon of memory.

TALKING TO YOUR FUTURE SELF

Ere the parting hour go by,
Quick, thy tablets, Memory!
—MATTHEW ARNOLD

In the movie *Memento,* Leonard Shelby suffers from an inability to convert short-term memory into long-term memory—a condition known as anterograde amnesia. He can remember what's happening in a five-minute window, but anything older than that fades away. As a result, he tattoos critical information directly onto his skin so he won't forget his mission. It's his way of talking to himself through time.

We're all like Leonard Shelby, but we etch the critical *where-have-I-been* information into our neural circuitry rather than onto our skin.

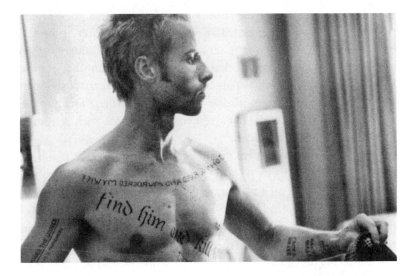

This is how our future selves know what we've been through, and thereby what to do next.

Almost twenty-four hundred years ago, Aristotle made a first attempt at describing this process, in his manuscript *De memoria et reminiscentia* (*On Memory and Reminiscence*). He used the analogy of pressing an imprint onto a wax seal. Unfortunately for Aristotle, he had no data to draw on, so the neural magic by which an event in the world becomes a memory in the head would remain enshrouded in mystery for millennia.

Neuroscience is just now beginning to unlock the puzzle. We know that when you learn a new fact—say, your new neighbor's name— there are physical changes in the structure of your brain. For decades, neuroscientists have slaved over laboratory benches to understand what those changes are, how they are orchestrated across vast seas of neurons, how they embody knowledge, and how they can be read out decades later. As a result, even while many puzzle pieces are missing, a picture is forming.

Simple forms of memory have been intensively studied at the cellular and network levels in modest organisms such as the sea slug. Why the sea slug? Its neurons are large and few, making it quite a bit easier to study than a human. Here's how a typical experiment goes: Scientists poke the sea slug gently with a stick. It withdraws. But if the scientists repeat this every ninety seconds, the sea slug eventually stops withdrawing. It "remembers" that there's nothing harmful about the stimulus. Now the scientists pair the stick with an electrical shock to the tail, and after that the withdrawal reflex from a mere touch of the stick becomes large: the sea slug "remembers" that the stick is teamed up with something dangerous.[2]

Such experiments have taught a great deal about changes that happen at a molecular level; however, animals that arrived later to the evolutionary party (such as mammals) have memory capacities that are much more powerful and far-reaching than those seen in invertebrates. Humans can remember details of our autobiographies. We can remember what we have dreamed and imagined. We can remember the spatial details of vast regions of geography. We can acquire complex skills that allow us to navigate commercial, social, and climatic conditions. Conveniently, we also have the ability to forget irrelevant minutiae, such as the location of an airport parking place two weeks ago, or the exact wording of a conversation.

The first systematic investigation into the physical basis of memory in mammals was undertaken in the 1920s by the Harvard neurobiologist Karl Lashley. He reasoned that if he could teach a rat something new (such as a route through a maze), then he might be able to erase that new memory by cutting into a little piece of the rat's brain in the right spot. All he had to do was find that magical spot, extract it, and demonstrate that the rat couldn't remember the route.

So he trained twenty rats to run a maze. Then he used his scalpel to cut through a different area of cortex in each animal. After giving them time to recover, he retested each rat to see which areas of damage had removed knowledge of the maze.

The experiment was a failure: all the rats were perfectly proficient at remembering the maze. None forgot the route.

The experiment's failure was its lasting success. Lashley realized that a rat's memory of the maze could not be localized to any single spot. Memory was not confined to a particular area but instead distributed broadly. The experiment revealed that there is no such thing as a dedicated memory structure in the brain. Memory storage is not like a filing cabinet, but instead like distributed cloud computing—similar to the way your email inbox is scattered on servers around the planet, often with high redundancy.

But how does a memory—such as a name, a skiing trip, a piece of music—get written into a widely distributed set of billions of cells? What is the programming language that translates from the realm of experience into the realm of the physical?

In the nineteenth century, before high-resolution microscopy, it was assumed that the nervous system, with its myriad fibrous highways coursing the body, was a continuous network like blood vessels. This view was not challenged until a century ago, when the Spanish neuroscientist Santiago Ramón y Cajal realized instead that the brain is a coalition of billions of discrete cells. Instead of a highway, the nervous system was more like a patchwork of local road projects that communicated with one another. He called his framework the "neuron doctrine," an insight that earned him one of the first Nobel prizes.

The neuron doctrine ushered in an important new question: If brain cells are separate, how do they communicate with each other? And the answer was quickly determined: they are connected at the specialized points we now call synapses. Ramón y Cajal suggested that learning and memory might occur by changes in the strengths of the synaptic connections.

By 1949, the neuroscientist Donald Hebb had been chewing on the idea and was able to refine it. He suggested that if cell A consistently participates in driving cell B, the connection between them will be strengthened ("potentiated").[3] In other words: fire together, wire together.

At the time Hebb proposed his hypothesis, there was no experimental evidence that could be marshaled for its support. Then, in 1973, two researchers discovered something that suggested Hebb might have nailed it. After stimulating input nerve fibers in an area called the hippocampus, they found an increased electrical response from the receiving (postsynaptic) cell. And that larger signal lasted for up to ten hours. They called this long-term potentiation, and it was the first demonstration that the strength of connections could be modified as a result of their recent history.[4]

Everyone quickly guessed the next step: what goes up needs the capacity to come back down. If a connection can potentiate, it also needs the ability to depress. Otherwise the network will saturate, becoming unable to store anything new. By the 1990s, it was shown that various manipulations (for example, A fires with no response from B) can lead to long-term depression; that is, the strength between two cells is weakened.

Scientists concluded they had found the physical basis of memory.[5] After all, subtly changing the strength of connections can radically change the output behavior of a network. Activity flows through the system based on what happened before. By tuning its parameters just right, a network can make links between things that occurred at the same time. The idea is that a simple mechanism like this could underlie all of the memories in your life.

Think of your best friend and your best friend's house. The sight of your pal triggers a particular constellation of neurons, and the house triggers another. Because the two groups of neurons are active at the same time when you go over there, the two concepts become associated; hence this is called associative learning. When either of these notions become triggered, it sparks the other to life. And even better, either one can activate all sorts of other associations, such as memories of your shared conversations, meals, and laughter.

Thus, in the early 1980s, the physicist John Hopfield tried to understand whether a very simplified artificial neural network could store a small collection of "memories."[6] He found that if he exposed a network to some patterns (such as letters of the alphabet) and strengthened the synapses between neurons that fired simultaneously, then the network would remember the patterns. Each letter (say, E) triggered a particular constellation of neurons, and these neurons strengthened their connections with one another. The letter S, in contrast, would be represented by a different pattern. Now Hopfield could present a corrupted version of one of the patterns (such as an E with part of the top cut off), and the cascade of activity running through the network would evolve toward the pattern for the full *E*. In other words, the network completed the pattern to match its notion of what an *E* should look like, given all of its previous experiences. Moreover, these networks were surprisingly robust to degradation: if you deleted a few of the nodes, the distributed memories of the network would still be retrievable. Hopfield had provided a powerful demonstration of memory in a simple artificial neural network, and it opened the door to a flurry of studies on "Hopfield nets."[7]

In the intervening decades, and especially in recent years, the field of artificial neural networks has taken off. In large part the rise of the field has been not because of new theoretical advances but instead because of massive computing power that allows giant artificial networks to be simulated with millions or billions of units.[8] Such networks have been able to do remarkable feats, such as trump the best chess players and Go players in the world.

But despite the fanfare, artificial neural networks are still a long way from operating the way the brain does. Although they are mind-blowingly impressive, they fail catastrophically when they are asked to switch tasks—say, from distinguishing cats and dogs to distinguishing birds from fish. Artificial neural networks are inspired by the brain, but they've gone off in their own simplified direction. To understand the magic in the brain (that is, what it can do that artificial neural networks so far cannot), we require a clearheaded look at the challenges and tricks of real, biological memory.

THE ENEMY OF MEMORY IS NOT TIME; IT'S OTHER MEMORIES

The first problem facing brains is that they live long lives. Animals are faced with changing, challenging environments, and therefore need to take on new information continually over years or decades. But lifelong learning must continually balance between the two horns of the bull: protect old data while taking on new. In artificial neural networks, learning is done in a "training phase" (typically with millions of examples) and then tested subsequently in the "recall" phase. Animals don't have that luxury. They have to learn and remember on the fly throughout a lifetime.

Unfortunately, memory models based on the basic textbook principles of synaptic change immediately collide with a problem: while Hebbian learning is great for encoding memory, it *continues* to be great for encoding memory, and things previously learned become quickly overwritten.[9] Artificial networks full of memory degrade into memory mud. Earlier memories are blurred out after new activity in the system, so quickly that you wouldn't be able to remember how a play started by the time you're at the end of act one. This problem is known as the stability/plasticity dilemma: How does the brain retain what it's learned while simultaneously taking on the new? Somehow, memories need to be protected. Not against the ravages of time, but against the invasion of other memories.

While artificial neural networks suffer from the memory mud problem, real brains don't. Reading a new book does not overwrite your spouse's name in your memory, nor does learning a new vocabulary word make the rest of your vocabulary slightly worse.

The fact that brains circumvent this dilemma, somehow locking down older memories, tells us that simply strengthening and weakening synapses in a network isn't the full picture. Something more is happening.

The first solution to the stability/plasticity dilemma is to make sure the whole system isn't changing at once. Instead, flexibility should turn on and off only in small spots, as steered by relevance. As we saw earlier, neuromodulators can carefully control the plasticity of synapses— and in this way, learning can take place only at the appropriate places and times, instead of each time activity passes through the network.[10] This specificity slows the descent of a network into memory mud, because it changes synaptic strengths only when something important is happening: you hear the name of a new colleague, a piece of news about a parent, or that a new season of your favorite television show is airing. But the network doesn't have to change when it regards a random street sign, or the color of a passerby's shirt, or the pattern of cracks in the sidewalk. This change-only-when-relevant feature reminds us that the brain is not simply a blank slate upon which the world scrawls all its stories. Instead, the brain comes pre-equipped for certain types of learning in particular types of situations. Experiences turn into memories when they are germane to the life of the organism, and especially when connected to a high emotional state such as fear or pleasure. This reduces the chances of overwhelming a network, because not everything gets written down.

But it doesn't *solve* the stability/plasticity problem, because there are still plenty of salient memories to worry about storing.

So the brain implements a second solution. It doesn't always hold memories in one place. Instead, it passes what it has learned to another area for more permanent storage.

PARTS OF THE BRAIN TEACH OTHER PARTS

Consider a warehouse. If you were constantly receiving new shipments of boxes, eventually the property would fill up. But if you're shipping the boxes off as they come in, you can maintain the space. In this way, memories often don't stay where they first formed, but get moved along.

Some of what we know about memory is distilled from data in the hippocampus and its surrounding regions, a central site of memory formation. In 1953, a twenty-seven-year-old patient named Henry Molaison underwent surgery to relieve his epilepsy—and for this purpose the hippocampus was removed on both sides of his brain. Postsurgically, Molaison was discovered to have a profound amnesia: he had lost his ability to form new memories or learn new facts. Surprisingly, he could still acquire a limited range of new skills (such as reading in a mirror), although he had no recollection of having acquired the skill. As detailed studies by Brenda Milner and her colleagues revealed, his memory for events that occurred before his surgery was close to normal. His case focused attention on the hippocampus, and specifically why it was crucial to *learning* facts but not critical for *remembering* facts that had already been learned.[11]

The answer? The role of the hippocampus in learning is temporary. It's not the site of permanent storage: Molaison could remember detailed autobiographical events from before the surgery.[12] The formation of new memories requires the hippocampus, but the memories are not stored permanently there. Instead, it passes along the learning to parts of the cortex, which hold the memory more permanently.

So how do the memories get from the way station of the hippocampus into their more permanent home in the cortex? One proposal is that stable storage cannot be achieved the first time a pattern of activity goes through the cortex; instead, an area such as the hippocampus must *reactivate* the trace several times to lock the memory into the cortex. This framework suggests why the hippocampus is necessary to consolidate memory: it needs to replay the patterns to the cortex over and over.[13] Once the memories are in the cortex, they gain stabilization with time. In Molaison's case: no rehearsal, no long-term storage. The system remains as it was before.

We see this movement of memory in many parts of the brain. Imagine you learn a new association: a red square means you should lift your arm, while a blue circle means you should clap your hands. You practice and you get faster. During this skill learning, changes are detectable quickly in certain brain regions (e.g., the caudate nucleus) that pick up on rewarded associations. However, if you keep performing the task, activity can eventually be detected in other areas (your prefrontal cortex). Those neurons are changing at a slower pace, suggesting that the first region is teaching the second what it has learned.[14]

As another example, when you learn how to rollerblade for the first time, you have to pay close attention to your limbs and invest major cognitive effort. But after many days of practice, you don't have to think about it anymore: it becomes automatized. This is because the parts of the brain involved in motor learning (the basal ganglia) pass the learning to parts such as the cerebellum.

The idea of shipping off the packages helps with the stability/plasticity dilemma, but there is still a problem of limited space. If you're shipping your boxes all over the world, there's no problem. But if you're just pushing the packages over to a different warehouse, you're simply kicking the problem down the road: the second warehouse will soon get filled up.

And that brings us to the trailhead of a third and deeper solution.

BEYOND SYNAPSES

The demonstrations of synaptic change have inspired thousands of researchers to chart out the detailed landscape of the phenomenon and unmask the molecular machinery that makes it possible. However, synaptic strengthening and weakening is not the only, or even the most important, mechanism involved in memory.[15] After decades of studying synaptic changes, we know that synaptic plasticity is necessary for learning and memory, but we have no evidence that it's sufficient. Perhaps changes in synaptic strength are simply the way intertwined cells carefully balance excitation against inhibition to keep themselves away from epilepsy (over-excitement) or shutdown (over-inhibition), and thus the synaptic changes are *consequences* of memory storage rather than the root mechanism. Although changes at individual synapses have received the most attention, both theoretically and experimentally, there are many other possible ways to store activity-dependent changes. By concentrating so intently on synaptic changes, the field may be missing part of memory's Rosetta stone. After all, everywhere we look in the nervous system, we find adjustable parameters. Nature has thousands of tricks to stockpile small alterations, all of which can change the behavior of a network.

Imagine you were an alien discovering human beings for the first time. You would be mystified at the number of moving pieces and parts that make up the flowing system we call the brain. As you watched humans interact throughout the day, your high-resolution eyes would see changes in the shapes of neurons, such as the growth or shrinkage of the dendrites based on experience. Squinting further into the system, you would observe changes in the amount of chemical messenger released by one cell for communication to another. You would detect changes in the number of receptors assembled to receive that chemical message. You would spot changes in the chemical decorations that hang off the receptors to change their function. You would be awed by the sophisticated cascades of molecules and ions inside the neurons, performing computations and adjusting themselves with every new input. In the neuron's nucleus, at the level of the genome, you would see ornate chemical structures attach themselves to winding strands of DNA, causing some genes to be expressed more while others are repressed.

You would likely be baffled by such a system, because plasticity is taking place within every one of these mechanisms. They are all flexible. Parameters change at all scales, from the growth and insertion of newly born neurons to changes in gene expression. With so many degrees of freedom in biological systems, the possibilities are vast for memory-storage strategies.

In fact, we have plenty of good reasons to think that synapses are not the only things changing. First, if learning only tuned the efficacies of existing synapses, we wouldn't expect large changes in the structure of the brain. But sizable changes can be seen in brain imaging when volunteers learn juggling, or medical students study for exams, or cabbies memorize the streets of London.[16] The cortical changes are more than the modification of synapses, but instead appear to involve the addition of new cellular material.[17]

Second, if memories were simply retained in the fabric of synaptic weights, we'd have no reason to expect *neurogenesis*: the growth and insertion of new neurons.[18] In fact, fresh neurons inserting themselves into the network would be expected to scramble up the delicate synaptic pattern. And yet there they are: a stream of new neurons being born in the hippocampus and trucking their way into the adult cortex. They're not accidental; they can be pinned to memory formation. For example, if you train a rat on a learning task that requires the hippocampus, the number of new adult-generated neurons doubles from baseline. In contrast, if you train rats on a learning task that doesn't require the hippocampus, the number of new cells goes unaltered.[19]

Third, alterations to the sugars and proteins around the DNA alter patterns of gene expression.[20] In this relatively new field called epigenetics, we find that world experience modifies which genes get squelched and which magnified. As an example, well-nurtured mouse pups (those receiving frequent licking and grooming from their mothers) show lifelong alterations in the patterns of the molecules that hang on to the strands of DNA, and this appears to decrease anxiety and increase nurturing in the offspring throughout their lives.[21] In this manner, your experiences with the world get under your skin—all the way to the level of your gene expression, where they can be embedded on a long timescale.

When neuroscientists and AI engineers speak about changes in a network, they're typically considering changes in the strength of connections between cells. But with your fresh alien eyes, it is clear why synapses are doomed to be insufficient: plasticity exists throughout the brain at every level. The way activity flows in networks depends on all the settings in the network, from large to small. Everywhere we probe, we find plasticity. So why does the field concentrate almost entirely on synapses? Because that's what we can most easily measure. The rest of the action is generally too tiny for our current technology during the fast-paced dynamics of a living brain. So, like a drunk looking for his keys under the streetlight, we concentrate mostly on what we're able to see.

So the brain has many knobs it can dial, and this leads us to the next part of the story: With all these possible parameters, how does the brain modify anything without muddling up function elsewhere? How can we understand the interaction of all the pieces and parts? What are the principles by which many degrees of freedom don't spin out of control, but instead keep each other in a system of checks and balances?

I propose that the most important lens is not what the biological parts are, but instead the *timescale* on which they operate. The story has to be told not in terms of the details of the mechanisms but instead the tempo at which they live.

DAISY-CHAINING A RANGE OF TIMESCALES

Some years ago, the writer Stewart Brand proposed that to understand a civilization, you need to look at multiple layers functioning simultaneously at different speeds.[22] Fashion changes rapidly, while the business pursuits in an area alter more slowly. Infrastructure—such as roads and buildings—evolves more gradually. The rules and laws of a society—the governance—adapt very slowly, wanting to fasten things down against the winds of change. Culture moves on an unhurried timetable of its own, resting on its deep foundations of story and tradition. At the slowest scale, nature plods at the pace of centuries or millennia.

Although it's not always noticed, all scales interact with one another. The faster layers instruct the slow layers with accrued innovations. The slower layers provide checks and structure to the fast layers. The power and resilience of a culture arise not from any one level of the system but from their interaction.

The principle of pace layering is useful for thinking about the brain. Instead of ranging from fashion to governance to nature, the brain's

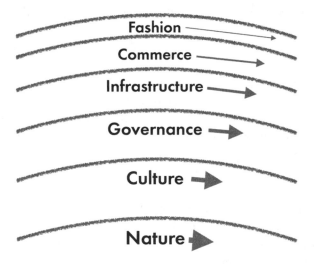

Pace layers.

pace layers range from fast biochemical cascades to changes in gene expression. Not only do synapses change, but so do many other parameters (for the connoisseurs, these include channel types, channel distributions, phosphorylation states, the shapes of neurites, the rate of ion transportation, nitric oxide production rates, biochemical cascades, spatial arrangements of enzymes, and gene expression). If such flows are linked up correctly, a transient event can leave a trace, because the fast cascades kick off slower cascades, which can eventually tip over more sluggish processes, which can set into motion gradual, deep changes. In this way, plastic changes are distributed along a time spectrum, not simply stored as all-or-none changes. All the forms of plasticity interact with one another, and the power of the system emerges from the layers operating in concert.[23]

We can see the results of this multi-paced system in many ways. Say you develop a crush on someone who's deaf. You work to learn sign language. Each time you successfully sign something, you get rewarded in the form of a flirtatious smile from your infatuation. You get quite good at sign language, almost fluent, and then suddenly your deaf fascination moves out of the country. You get no more reward from making the signs with your lonely fingers. With no returns on effort, you eventually forget how to speak sign language. It seems that the story ends there. But three years later a new deaf person moves to town. Possibly because of a wistful nostalgia, you find this one equally attractive, and so you try your hand at sign language again. Alas, you've forgotten your sign language entirely: your fingers just can't remember what to do. You lament, because last time it took you two months to get good at sign language, and you're certain your new fixation doesn't possess that kind of patience. But you discover that this time the learning is much faster. In fact, within three days you're flirting fluently. Even though you were certain it was all forgotten, there you are, signing like a pro.

The time savings between the first and the second times means that something in your brain held on to that information, even during those desolate years with no practice.[24] The savings result from slow changes in the deeper parts of the system. During your first infatuation, fast-moving parts learned the task, and with increasing practice passed the changes on to the deeper layers. When your crush left on a jet plane, the faster layers quickly adjusted their behavior. But deeper parts were hesitant to follow—reluctant to abandon the long, slow learning they had invested. Thus, when the next deaf splendor arrived, the deeper layers found themselves already prepared for sign language, and this resulted in savings. A skill you thought was gone was still present, burned deep into the circuitry.

Hidden savings in the brain are found in many settings, including outer space. When an astronaut returns from a long journey in orbit, she doesn't step out of the capsule and walk over to Starbucks; instead, she has to remember how to walk in earth's gravity, almost as though she were learning afresh. But she relearns quickly; she doesn't have to recapitulate infancy. In fact, her performance immediately after the flight reveals the depth of the savings in the brain, and therefore gives a good prediction of how quickly she'll be walking again.[25]

A sense of the brain's pace layers also shines light on the concept of schema that we learned earlier. Remember Destin and the trick bicycle? I mentioned that after months of learning how to ride with the reversed handlebars, he found himself unable to ride a normal bike. But this confusion didn't last long, and soon he could switch bikes easily. He had a schema for each. And we can now understand schema at a deeper level. It is not that short-term learnings overwrite one another (*I've learned how to ride the reversed bicycle, and now the program for a normal bicycle is gone*). Instead, the two programs live in deep layers. After his training, Destin has burned both programs into long-term circuitry, and the context (*which bike am I on?*) steers the correct path through the network.

In the end, exceptionally useful programs get burned down all the way to the level of the DNA. Consider instincts—the inborn behaviors we don't have to learn.[26] These come about via plasticity on a longer timescale: the Darwinian plasticity of species. By natural selection over millennia, those with instincts that favor survival and reproduction tend to multiply.

A century ago, one of the challenges to understanding memory was the lack of technology. Now one of the challenges is the presence of technology—especially computers. The digital revolution has so thoroughly changed every aspect of our lives that it is sometimes hard to shake metaphors, even when they are badly mismatched. This is nowhere more apparent than with the word "memory." Human brains do not store memories the way computers do. Instead, brains retain and retrieve the memory of a movie without encoding it pixel by pixel, and we remember and reproduce our favorite stories without encoding them word for word. When someone tells you a joke, for instance, you do not encode a neural log file of each word and its inflection. Instead, you understand the *gist* of the joke. If you are bilingual, you may hear the joke in one language and turn around to tell it to someone else in a different language. The joke is not about the exact words but instead about concepts triggered internally.

Instead of encoding pixels or transcripts, we encode new stimuli with respect to other things we have learned, including concepts both physical and social. What we learn is represented in terms of what we already know. Two people can look at a list of important dates in Mongolian history, but if one of them commands a richly developed model of Mongolia, the new facts are more readily incorporated into her network of knowledge. For the other, who knows little about the country and has never been there, the facts have little scaffolding upon which to attach.

Recall that in the pace layers model, slow layers provide a framework for the fast layers. As a result, early experience becomes foundational. It develops into the architecture upon which everything subsequent is built. Everything new is understood through the filter of the old.

For better or worse, this makes some dreams of the future impossible. In the movie *The Matrix*, Neo and Trinity come across a B-212 helicopter on the top of a building. Neo asks, "Can you fly that helicopter?" Trinity replies, "Not yet," rings her colleague, and asks for "a pilot program for a B-212 helicopter." Her colleague frenetically hits keys on a collection of computers, and within a few seconds the pro-

gram is uploaded into Trinity's brain. Neo and Trinity board the helicopter, and she expertly steers it between buildings.

We would all love this future, but it's not going to happen. Why not?
Because memory is a function of everything that has come before it.
One person's knowledge of flying a B-212 helicopter might be encoded
by its similarity to riding a motorcycle. Another person might have
grown up on horses, so she builds the piloting knowledge on top of
motor memories of steering a steed. A third person stores the knowledge in the context of a childhood video game. Each person grasps
the task differently, making it impossible to have a standard set of
instructions that is uploadable to any brain. In other words, unlike a
computer, the "instructions" for flying the machine aren't a file; they
are instead tied to everything that has come before in your life. Earlier
experiences build an internal city of memory, into which each new
resident must find his unique fit.[27]

The key thing to understand about pace layers is the interaction
between them. As the field of neuroscience marches on, I suspect
that many clinical issues will come to be understood in terms of such
interactions.

For example, recall Lord Admiral Nelson: After he was shot by a
musket, his arm was amputated, but he spent his remaining years feeling as though his absent arm were still somehow present. Although the
cortex that formerly responded to touch on the arm became reactive
to touch on his face, downstream brain areas still expected that patch
of cortex to represent his arm. In other words, to the slow, deep layers,
activity in that patch continued to be interpreted as arm sensation. As
is typical in amputees, that led to perceptual confusion in the form of
a phantom sensation: he was certain his arm still existed, for the deep
layers told him so. The pace layer system works best for things changing at normal speeds—but an upheaval of the body's design can pitch
the system into a strange state, especially when change arrives at the
speed of a musket ball.

As another example, consider an unusual condition called hyperthymesia, in which a person has an essentially perfect autobiographi-

cal memory: she forgets almost nothing. Name any date in her past, and she can tell you the weather that day, what she did, what she wore, and whom she saw. When the field of neuroscience possesses the technology to get to the bottom of this phenomenon (at the neuronal and molecular levels), it will almost certainly be understood as an interaction between the layers, such as the interfacing of the layers at an unusual speed. In terms of a society, it would be as if the fashionistas obtained too much power and pushed their latest fads directly into the governance layer. (As a side note, although it might sound great to remember everything, hyperthymestics suffer with the inability to forget the trivial. As Honoré de Balzac once said, "Memories beautify life, but only forgetting makes it bearable.")

Finally, consider synesthesia, a condition in which stimulation of one sense triggers automatic, involuntary experiences in a second pathway. For example, a letter of the alphabet produces an internal color experience—such as *J* triggering an internal sensation of purple, or *W* eliciting green.

The most common hypothesis is that synesthesia reflects an increased degree of cross talk between normally separated brain areas. But I have previously suggested a different hypothesis: that it represents "sticky plasticity."[28] Let's imagine that a young child sees a purple *J*—perhaps as a sign on an elementary school wall, or sewn on a quilt, or as a self-made choice from a box of crayons. As we saw, synapses can modify their strength if their neurons are active at the same time—say, those coding for *J* and those for purple. They fire together, and so they wire together. Now, for most people, the connection between *J* and some color will continue to be modified with each new sighting of the letter *J* in different hues. Thus, when a yellow *J* is seen, the connection between *J* and yellow is strengthened and the connection between *J* and purple is weakened. With enough exposure to differently colored *J*s, the letter-color pairings will average out, leaving no particular association between letters and colors. I suggest that synesthetes have atypical plasticity; specifically, a reduced ability to modify an association once it has been set. Once an initial pairing between a letter and a color has been set, it *sticks*.

How could this be tested? After all, when you look at one synesthete's alphabet colors, it generally looks quite different from another's. So how would you ever know if they had imprinted on something they had seen as children?

To test this hypothesis, I built the Synesthesia Battery,[29] an online assessment to verify and quantify synesthesia. I collected and verified data from thousands of participants, and with two of my colleagues from Stanford I carefully analyzed the colored alphabets from 6,588 synesthetes. What we found came as a great surprise. Although the mapping of letters to colors was essentially random among most participants, there were also hundreds of synesthetes who all had approximately the same pattern: *A* was red, *B* was orange, *C* was yellow, *D* was green, *E* was blue, *F* was purple, and then the cycle repeated again with a red *G*.[30] Stranger still, all the synesthetes with this particular pattern were born between the late 1960s and the late 1980s. Within that window, upward of 15 percent of synesthetes had this same letter-

A	red	N	orange
B	orange	O	yellow
C	yellow	P	green
D	green	Q	blue
E	blue	R	purple
F	purple	S	red
G	red	T	orange
H	orange	U	yellow
I	yellow	V	green
J	green	W	blue
K	blue	X	purple
L	purple	Y	red
M	red	Z	orange

Many synesthetes born between the late 1960s and the late 1980s
perceive alphabets that match the colors of the Fisher-Price refrigerator magnet set.
One of our participants had photographic evidence that he had received the
Fisher-Price set as a child.

color relationship. None of the synesthetes born before 1967 had this pattern; nor did almost anyone born after the 1990s.

The colors turned out to be those of the Fisher-Price magnet set, which was produced only between 1971 and 1990 and adorned refrigerators all over America. The magnet set didn't cause synesthesia; instead, for people who were predisposed to it, the magnets became the source of the letter-color pairings.[31]

Synesthesia, like hyperthymesia, reflects a stickiness in pace layering: the fast layers push their agendas more quickly than normal into the deeper layers. While hyperthymesia and synesthesia are not considered diseases, they are statistically unusual—which suggests that the speed of interaction between the neural pace layers in the majority of the population has been evolutionarily optimized.

MANY KINDS OF MEMORY

While talking about memory in this chapter, we've spoken as though it were one thing. But memory has many faces.

Consider a case like Jody Roberts, who in 1985 worked in Washington State as a journalist. One day, she vanished. Her loved ones searched assiduously and after many years resigned themselves to the tragic conclusion that she was dead.

But she wasn't. Five days after her disappearance, she showed up a thousand miles away, wandering disoriented in a shopping mall in Aurora, Colorado. She had no identification on her, just the key to a car that was never found. She had total amnesia about her past. Police took her to the hospital. Jody wasn't able to figure out her identity, so she took on the name Jane Dee, began work at a fast-food restaurant, and enrolled herself at the University of Denver. Eventually she moved to Alaska, where she married a fisherman, got a job as a web designer, and mothered two sets of twins.

Twelve years later, an acquaintance recognized Jody from a news story. Jody was reunited with her sobbing and thankful family. But she had no memory of them. She was polite but distant. As her father stated to the news, "She is the same basic person. We got her back, in a sense."[32]

The key thing to note about stories like Jody's is that she could still remember how to speak English, how to drive, how to flirt, how to get a job, how to waitress, how to write love letters, and how to take care of children. She just couldn't remember her autobiography. Cases like Jody's (there are many of them) lead to the realization that there are many kinds of memory. Contrary to first glance, memory is not one thing, but instead comprises many different subtypes. On the broadest level, there is short-term memory (remembering a phone number just long enough to dial it) and long-term memory (what you did on that vacation two years ago). Within long-term memory, we can distinguish declarative memories (such as names and facts) from nondeclarative memories (how to ride a bicycle, something you can *do* but not articulate how). Within the non-declarative category are several subtypes, such as remembering how to type rapidly or why you salivate when hearing someone open a candy wrapper.

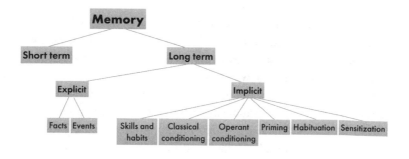

Different kinds of memory.

The first step to understanding Jody's situation is recognizing that different brain structures support different kinds of learning and memory. Injury to the hippocampus and surrounding structures affects the formation of new declarative memories (*what did I eat for breakfast this morning?*), but not non-declarative memories (how to speak, sing, walk). This is why the amnesiac Henry Molaison could function well in his daily life, with no deficit in brushing his teeth, driving his car, or having a conversation. Other brain areas are required for learning motor skills, especially those involving balance and coordination. Other areas are important for linking motor acts with subsequent rewards. Other areas are critical in memory changes related to fear conditioning, and various reward structures support learning successful foraging strategies. The list of brain structures and their relationship to learning and memory is large and increasing, and Jody and Henry teach us that the integrity of a particular subsystem is not necessarily essential to the function of others. You can lose the ability to remember the narrative of your life, but this has no bearing on your ability to learn and remember new motor skills.

Consider this example: from your childhood you've seen lots of birds, and so your brain made the generalization that animals with feathers can fly. But you've also seen ostriches at the zoo, and you were able to retain that particular exception to the rule. As well, you might learn that the ostrich in your zoo is named Dora, which does not apply to other ostriches you might encounter.

Some years ago, people building artificial neural networks began to run into a problem with this distinction between generalizations and specific examples. They could build networks that learned generalizations (*things with feathers fly*), or they could build a network that held a collection of specific examples (*the bird named Dora does not fly, while the one named Paul does fly*). But they couldn't do both. Either the network changed its parameters slowly by being exposed to thousands of examples, or it changed things quickly, pushed around by single examples.

How do you get a single brain to learn things at both timescales at once? After all, you need different timescales of learning to remember different sorts of facts about the world. Sometimes you want to generalize (*lemons are yellow*), and other times you need to remember something specific (*the lemon in the veggie drawer of my fridge is rotten*).

This apparent incompatibility of goals yielded an important clue.[33] To do both jobs well, the brain has to have different systems with different speeds of learning: one for the extraction of generalities in the environment (slow learning), and one for episodic memory (fast learning). One proposal is that these two systems are the hippocampus and the cortex: the hippocampus is fast in its changes (so it learns from examples rapidly), while the cortex takes its time to slowly extract generalities. The first changes quickly, retaining particulars, while the second changes slowly, requiring many examples. With this trick, the brain can do rapid learning from individual episodes (*this button makes the rental car start*), and at the same time it can be running a slow extraction of statistics across experiences (*most flowers bloom in the spring*).[34]

MODIFIED BY HISTORY

As activity passes through the brain, it changes the structure. From the point of view of the vast forest of neurons in your skull, the organizational problem is tremendous: the nervous system must physically change itself to optimally reflect the world in which it's embedded. The individual changes must each make the right contribution to the network to embody the new knowledge, and the changes must be positioned to make a difference to behavior when the right moment arises, some time in the future. A simplifying error when thinking about memory has been to assume that it is underpinned by a single mechanism of change. The classical story of strengthening and weakening synapses has brought us a long way, and artificial neural networks employing those principles can perform impressive engineering feats. But memory is more than dialing synapses in a large connection diagram. As we saw, simple synaptic models quickly lose the capacity to represent old data as new data stream in. The way memories degrade—with older memories having more stability—reveals the secret of different timescales of change.

The synaptic model would be convenient for neuroscientists and

AI engineers, but it's almost certainly not nature's approach. Instead, the changes underlying memory are distributed widely over titanic numbers of neurons, synapses, molecules, and genes. By analogy, just consider how the desert remembers the wind: it does so in the slope of its sand dunes, in the shape of its rocks, and in the evolutionary pressures that carve the wings of its insects and the leaves of its plants.

Progress in the memory field calls for a maximally realistic view of the phenomenon we are attempting to explain. Although current artificial neural networks succeed at wonderful feats (such as discriminating photographs with superhuman skill), they do not capture the basic characteristics of human memory. The richness of our memory, I suggest, emerges through a biological cascade of timescales. New information builds upon the old, fitting into the constraints offered by previous experience. I've known many medical students who worry that if they learn one more fact, something else will drop out of their memory. Happily, this constant volume model isn't true. Instead, with each new thing you learn, the better you're able to absorb the next related fact.

11

THE WOLF AND THE MARS ROVER

I recently read about a school in California that shut down its programs of arts, music, and physical education. Why all the budget trimming? Because some years earlier they had decided to funnel all the money toward a state-of-the-art computer center for their students. They purchased $330 million worth of computers, servers, monitors, and peripherals. With pomp and circumstance, they proudly unveiled their educational showpiece.

A few years later, the computer equipment had become outdated. Chips were faster, memory had moved from hard drives to the cloud, and new software was incompatible with the old firmware. Less than a decade after the initial purchase, they were forced to dump all their gear. The apple of the school's eye—the expenditure that had assassinated the creative arts and physical fitness—had lived its short lifetime and was now an expensive memory glinting in the landfill.

The story got me wondering. Why are we still building hardwired machines that end up discarded? The moment we solder circuitry into place, we immediately doom it with an expiration date.

If we are astute students of the biology surrounding us, we can capitalize on the principles of liveware. Consider that when a wolf gets a leg caught in a trap, it gnaws off its leg and continues with a limp. Contrast that with the Mars rover *Spirit*. It touched down on the surface of the red planet on January 4, 2004, and rolled around success-

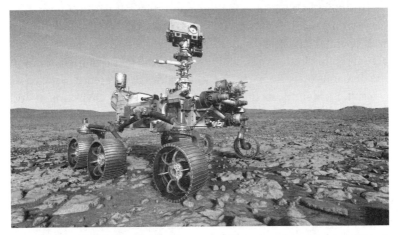

Spirit, *a wonderful rover that is now a $400 million piece of extraplanetary litter.*

fully for years. But then, in late 2009, the four-hundred-pound robotic vehicle became stuck in the soil. It couldn't get out, in part because its right front wheel had stopped working. Stuck in the Martian terrain, *Spirit*'s solar panels found themselves unable to orient toward the sun. The rover lost power and then sustained irrevocable damage during the winter. On March 22, 2010, it beamed its swan song to Earth and perished.

Spirit heroically outlasted its scheduled life span. But had we sent up human colonies that lasted only a few years before falling into a heap of bones, we'd be distraught.

This is no criticism of the astonishing engineering from NASA. The problem is that we're still building robots with hardwiring. If a current robot loses a wheel, an axle, or part of its motherboard, the game is over. But across the animal kingdom, organisms damage themselves and keep going. They limp, they drag, they hop, they favor a weakness, they do whatever it takes to keep moving in the direction of their goals.

The trapped wolf chews off its leg, and its brain adjusts to the unusual body plan—because getting back to safety is *relevant* to its reward systems. It needs food, shelter, and the support of the rest of the pack. So its brain figures out a solution to get there.

This difference between the rover and the wolf lies in information versus information-with-a-purpose. Unlike *Spirit,* the leg-trapped wolf operates with ambitions: to escape danger and to reach safety. Its actions and intentions are undergirded by the threat of predators and the demands of its stomach. The wolf traffics in deference to goals. As a result, its brain drinks up information about the environment and what its limbs allow it to do. And its brain translates those capabilities into the most useful actions.

The wolf carries on with a limp because animals don't shut down with moderate damage. And neither should our machines.

Mother Nature knows there is no point in hardwiring a wolf brain. Body plans change. Environments change. The complex relationship among capabilities and actions changes. Instead of predefined circuitry, the better plan is to build an infotropic system that optimizes everything on the fly, self-adjusting to become efficient at reaching its goals. Some purposes are long term (such as survival), while others are short term (working out pincer movements to catch escaping caribou); in all cases, the brain adjusts itself to target those purposes.

What would our robots require to keep going after they get damaged? An ability to drive a modified body plan, coupled with a yen to eat, socialize, survive. With that in place, they could lose wheels and damage parts, and their remaining circuitry would adjust to finish what they began. Just imagine the Mars rover sawing off its stuck wheel and figuring out how to shift mobility to its remaining wheels. Such principles could be used to build reconfiguring machines that combine input and goals to adapt their own wiring. When they lose tires, crush axles, or rip wiring, their remaining circuitry will reconstitute as necessary to finish what it began.

Just as there's no point hardwiring a wolf, there's no point hardwiring the Polgár sisters or Itzhak Perlman or Serena Williams. The world is too complex to foresee, and it would prove impossible to program genes to match the world's intricacy. After all, everything is in flux: bodies, food sources, and the mapping between inputs, capabilities, and outputs. Instead of predefined circuitry, a better approach is to build a system that actively improves, self-adjusting to reach its goals.

For decades, neuroscience has been fueled by the contributions of engineering, from oscilloscopes to electrodes to magnetic resonance imagers. The time might have finally arrived to reverse the direction of influence, allowing engineering to draft off biology.

With our current engineering in the fanciest clean rooms in the richest companies, we cannot come close to approaching what we see around us: mobile creatures from dogs to dolphins, humans to hummingbirds, pandas to pangolins. These creatures don't need to be plugged into wall sockets; they find their own energy sources. They climb, run, scale, jump, swim, crawl, and with a little effort can master skateboards, surfboards, and snowboards. This is all possible because Mother Nature endlessly plays with genes to build new sensors and muscles and the brain figures out how to take advantage of them. And these creatures can sustain damage—from a broken leg to a hemispherectomy—and they keep on trucking. Our devices have neither the flexibility nor the hardiness that characterizes biology.

So why haven't we built livewired devices yet? Let's not be too hard on ourselves: Mother Nature has had billions of years to test trillions of experiments in parallel. We can hardly imagine time vistas that size, or conceive the countless creature brains that have come to term and ambled on the earth or twirled in its waters or glided in its skies.

It's going to take us a while to catch up. The good news is that we're beginning to crack the codes around us.

So how can we start incorporating the principles of livewiring more deeply into what we build? The first answer is to imitate what Mother Nature has already developed. Take as an example the sensors that cover the body of a blind cave fish called the Mexican tetra: by detecting water pressure and flow, it can decipher the structures in the pitch-black water around it. Inspired by this, engineers in Singapore have built artificial versions of these sensors for submarines.[1] After all, lights on underwater craft are energy hungry and disruptive to ecosystems. Through the use of an array of small, low-power sensors inspired by the Mexican tetra, the hope is to "see" in blackness via the shifts of water.

While sensor biomimicry is a great start, it's only the beginning. The larger challenge is to design a nervous system that integrates new plug-and-play devices. Why would this be useful? Take as an example the problems that NASA continually faces with the International Space Station. Collaboration between nations is the heart of the project, but it's also the core of an engineering problem. The Russians build one module, the Americans attach another, the Chinese contribute another. The ISS faces a continual problem coordinating the sensors on the modules from the different countries. The American heat sensors don't always sync up with the Russian vibration sensors, and the Chinese gas sensors have trouble communicating to the rest of the station. The ISS continually throws engineers at the problem to solve and re-solve it.

The right way to tackle this, once and for all, would be to imitate Mother Nature. After all, she has booted up thousands of new sensors, from eyes to ears to noses to pressure sensors to heat pits to electroreceptors to magnetoreceptors and more. Over evolutionary vistas, she's invested her efforts into designing a nervous system that can extract the information of these sensors without having to be told anything about them (chapter 4). The sensors can be totally different in their design, and yet they have no difficulty working together seamlessly. Why? Because the brain moves around in the world, looks for correlations between the different incoming data streams, and figures out how to put incoming information to work.

How could we take advantage of this approach? One of the brain's most powerful techniques is to put out a motor act and assess the feedback. I suggest we should let the ISS experiment not only with its sensorium, but also its *motorium*—that is, how it uses its body. After all, the principle of the ISS is one of modular design, which means its body plan will change all the time. As we saw in chapter 5, brains learn to drive whatever body they find themselves in. No preprogramming is necessary, just motor babbling: trying out various moves and watching the results. In this way, brains figure out their bodies. By the same technique, the ISS could sporadically fidget and move about to figure out its new attachments and all its attendant capabilities. The future of self-configuration means that we will design machines that are not finished, but instead use interaction with the world to complete the patterns of their own wiring.

Once inbound and outgoing signals are coordinated, all kinds of magic can happen. As an example, take a popular kind of microchip that sits at the heart of many products (the field-programmable gate array, or FPGA). It's an amazing chip, but one of the main difficulties in these chips is coordinating the timing of all the signals dashing around inside it. Zeros and ones race around in the chips at close to the speed of light, and if a bit from one part of the chip accidentally arrives somewhere before a bit from another part, it's disastrous: the whole logical function of the chip is compromised. Timing in micro-chips is a whole subfield; there are thick books on the topic.[2]

Through the lens of a biologist, there is a simple solution. The brain faces the same challenge that a chip does: it deals with a constant stream of signals coming in (from sensory devices and internal organs) and a stream of signals going out (movements of the limbs). And the timing matters a great deal. If you believe you heard a twig crack just before your foot hit the ground, you'd better scan for predators. But if the crack happens just after your footfall, that's a normal sensory consequence of your own actions, calling for no panic. The challenge for the brain is that there's no way to preprogram the expected timings of the individual senses, because their timings can change. When you enter a dark place from a bright one, the speed at which your eyes talk to your brain slows by almost a tenth of a second. When it's hot instead of cold, signals can travel along your limbs at a faster speed. When you grow from an infant to an adult, the length of your limbs changes, and thus the amount of time to send and receive signals lengthens.

So how does the brain solve these timing problems? Not by read-ing a thick book on timing verification. It puts out probes into the world: kicking things, touching things, knocking on things. It operates under the assumption that if you *created* the action (by reaching out into the world), then the time-scattered information returning via the sensory channels should be perceived as synchronized. That is, your consciousness should adjust to see, hear, and feel the consequences all at the same time.³ After all, the best way to predict the future is to create it. Each time your brain interacts with the world, it sends a clear message to the different senses: synchronize your watches.

So the neuro-inspired way to solve the microchip timing problem would be to have the chip send probes to itself regularly (just the way a person might bounce a ball, tap silverware together, or look back and forth after putting on glasses). When the chip is the one "creating" the probe, it can have clear expectations about what is supposed to happen. And then it can adjust itself accordingly, allowing us to finally jettison the outsized books.

———————

As we incorporate the principles of livewiring into our machines, every type of device will come under our purview. Consider self-driving cars. In the future, we can presumably look forward to roadways with less carnage—not only because cars will have shared knowledge and communication with the surrounding cars, but also because there will be learning in the system: the cars will become better drivers with time. It's not that they'll be purposely programmed to make mistakes at the outset; instead, the problem is that the world is complex. Not all situations can be preprogrammed. So, like adolescents who learn from their mistakes and share their lessons, the cars will grow smarter with time.

We can also use principles of livewiring to distribute electricity with much greater efficiency than we do now. As we build out the Internet of Things (the connection of everyday devices to the net), we can push and pull resources from our colossal constellations of lights, air conditioners, and computers—using the internet as a giant nervous system that distributes electricity where and when it's needed.⁴ Among other things, such a smart grid would open the door to private power generation: think of adding windmills and solar farms the way Mother Nature adds new peripheral devices to a creature and lets its brain figure out how to use them. Beyond increasing efficiency, a smart grid could be able to withstand attacks by healing itself. Most countries of the world claim to be working to implement versions of smart grids, but the truth is that there are various levels represented by the word "smart." Your third grader is smart and Albert Einstein is smart. We'll slowly transition from a smart grid to a genius grid as we come to understand and implement the principles of livewiring that Mother Nature has conceived over billions of years.

Beyond the advantages of livewiring for rovers, cars, chips, and grids, I hope to see biology redefine fields like architecture. Currently, our most magnificent constructions pale in comparison to the creations of nature—from the handsome structure of a neuron to the exquisite design of the cerebellum to the limber dance of limbs. What if architects took inspiration from biology?

Imagine a building that senses the traffic through its bathrooms and releases attractants or repellents to call for the quick growth of more sink spigots, urinals, and sewage pipes. Or imagine a house that knows its own architecture and can readjust its nervous system to match changes: when a new room is added, air ducts and electrical wiring grow into it naturally. The brain of the house readjusts, developing a new sense of what the house looks like. Similarly, when part of the house is destroyed in an accident, resources are dynamically reconfigured: a damaged kitchen reallocates counter space and electronics to accomplish the same functions of the larger kitchen in a smaller area. We might have to deal with phantom refrigerator pain in the future, but at least we won't have to deal with the old-fashioned kind of house in which falling walls stop the show. And what if we engineered bricks that took cues from each other to self-organize into a structure—the way that the individual neurons assemble into larger nuclei? What if buildings could shift around, dynamically optimizing their sun exposure, their shade, their water access, and the amount of wind they were exposed to? What if they were mobile, able to stand up and move to a better spot when a fire looms or coastlines change over long timescales? There's no end to the way that engineering will flourish as we come to understand livewiring.

Finally, note that a future of self-configuring devices will change what it means to fix them. Construction workers or car mechanics are rarely surprised: the breaking of one part of the building or engine leads to a reasonably predictable set of consequences. In contrast, young neurologists are often uncertain and insecure. Although they can come to recognize and diagnose brain problems with reasonable accuracy, there is a frustration that patients don't typically fit the textbook models. Why do the textbooks fall short? Because each brain has followed a unique trajectory based on its history, goals, and practice. Construction workers and car mechanics in the distant future will have to be more like neurologists, feeling around for general principles rather than expecting to fish out a specific wire or bolt.

As we illuminate the principles of brain function, they will be gainfully applied to fields from AI to architecture, from microchips to Mars rovers. We won't have to keep filling dumps with brittle devices forevermore. Instead, self-reconfiguring devices will populate not only our biological world but also our manufactured world.

I suspect our distant descendants will look back on the history of the Industrial Revolution and wonder why it took us this long to level up simply by mimicking the principles of nature's billions-of-years-old biological revolution, one that surrounds us on all sides.

So when a young person asks you what our technology will look like in fifty years, you can reply, "The answer is right in back of your eyes."

12

FINDING ÖTZI'S LONG-LOST LOVE

In September 1991, a German couple hiking in the Tyrolean Alps came across a dead body. The bottom 90 percent of the body was frozen solidly into the glacial ice; only the head and shoulders were exposed. The man in the ice was perfectly intact and freeze-dried. The bodies of several wayward mountain climbers had been found in these mountains over the years, but this discovery was different.

This man froze here five thousand years ago.

The frozen specimen came to be known as the Tyrolean Iceman and was given the name Ötzi. He was ice-picked out of his prison over the course of several visits, was then refrozen by inclement weather, and was finally hacked out with ski poles. After several weeks of ownership debates by different jurisdictions, scientists were able to squeeze in and determine that the man hailed from the Late Neolithic period—specifically, the Copper Age.[1]

Question marks immediately began to sprout. Who was this man? What had he been like? In which regions had he traveled? As I read the scientific outpouring, I was amazed by how much could be gleaned from his simple remains. The contents of his gut revealed his last two meals (chamois meat and deer meat, both eaten with einkorn wheat bran, roots, and fruits). The pollen in his final meal had been fresh, placing his death in the spring. His hair told the general outline of his diet from the previous few months, and the copper particles in its strands suggested he was involved in smelting. The composition of his tooth enamel showed the region where he had spent his childhood. His blackened lungs told of the smoke of campfires. The proportions of his leg bones disclosed that he had spent his young life trekking great distances over mountainous regions. He had received primitive acupuncture for wear and tear in his knees, as told by the condition of the bones and corresponding cruciform marks on his skin. His fingernails chronicled his history of disease: three lines across the nails signified he had suffered from systemic illness on three occasions in the half year before he died.

One can gather a tremendous amount of data from a body, because a body is shaped by its experiences.

As we've seen, a much more specific shaping takes place in the brain.

At some point we might perhaps be able to read the rough details of someone's life—what he did and what was important to him—from the exact molding of his neural resources. If feasible, this would amount to a new kind of science. By looking at how the brain shaped itself, could we know what a person was exposed to, and perhaps what he cared about? Which hand did this person use for fine motor skills? What were the relevant signals in his environment? What was the structure of his language? And all the rest of the questions that we cannot answer by looking simply at guts and hair and knees and fingernails.

After all, this is the same logic with which we reverse engineer the shot-down warplanes of our enemies. We assume that function is related to the structure: if cockpit wires are in a particular configuration, there's a reason behind it. The same opportunities suggest themselves for retrospective brain decoding.

If all goes well, fifty years from now we'll revisit the glass-encased Tyrolean Iceman in Bolzano, Italy. We'll break him out from the clear prison that mirrors his glacier. And we'll read the details of his narrative etched directly into the fabric of his brain. We will understand his life not from the outside but from his own point of view. What mattered to him? What did he spend his time on? Whom did he love? Right now this is science fiction, but in a few decades it might be science.

We already know that evolution, on long timescales, carves creatures to match their environment. Consider the fact that the photoreceptors in our retinas are perfectly matched to the light spectrum given off by the sun, or that our genomes hold an archaeological record of ancient infections. But on the short timescale of a life, brain wiring can tell us so much more. The structure of the brain illuminates the concerns, time investments, and informational hotspots of a person's local environment. In this way, we not only could come to know the Tyrolean Iceman as a representative of his era, but could read the microscopic diary etched in the script of his brain cells. We could witness the faces of his siblings, his children, his elders, his friends, and his competitors; smell his rainy nights and his campfires; hear his language and the voices he knew; experience his personal joys, fears, heartbreaks, and hopes.

Ötzi was not lucky enough to live at a time when he could point a video camera at his world to capture it. But he didn't need to. He was the video camera.

WE HAVE MET THE SHAPE-SHIFTERS, AND THEY ARE US

People sometimes say things like this to me: "The doctors told my niece she would never walk again. And look at her now, jogging past!" First, I am thrilled for the patient and the family that it worked out. Second, I am a tad skeptical that their physician actually said "never." At least not without prefacing it with something like "the most *likely* scenario is." Or perhaps the physician was just trying to avoid a lawsuit by setting expectations low so any progress would be appreciated. Whatever the reason, a good physician would rarely commit to finality on such a statement, because the brain's ability to reconfigure keeps open the doors of possibility, especially in the young.

To my eyes, livewiring is quite possibly the most gorgeous phenomenon in biology. In this book I've endeavored to distill the main features of livewiring into seven principles:

1. *Reflect the world.* Brains match themselves to their input.
2. *Wrap around the inputs.* Brains leverage whatever information streams in.
3. *Drive any machinery.* Brains learn to control whatever body plan they discover themselves inside of.
4. *Retain what matters.* Brains distribute their resources based on relevance.
5. *Lock down stable information.* Some parts of the brain are more flexible than others, depending on the input.
6. *Compete or die.* Plasticity emerges from a struggle for survival of the parts of the system.
7. *Move toward the data.* The brain builds an internal model of the world, and adjusts whenever predictions are incorrect.

Livewiring is more than a jaw-dropping curiosity of nature; it is the fundamental trick that allows for memory, flexible intelligence, and civilizations. It is about finding oneself without the tools for a job and fine-tuning the brain to create those tools. Livewiring is the mechanism through which evolution by natural selection is relieved of some impossible pressures: instead of presaging every eventuality, brains can adjust billions of parameters on the fly to meet the unforeseen.

Plasticity is found at all levels, from synapses to whole brain regions. The constant fight for territory in the brain is a survival-of-the-fittest competition: each synapse, each neuron, each population, is fighting for resources. As the border wars are fought, the maps shift in such a way that the goals most important to the organism are always reflected in the structure of the brain.

Livewiring will become a standard part of our thinking: as we study the world around us, we'll see with increasing clarity the brain's role.

Consider the precipitous drop in American crime in the mid-1990s. One hypothesis is that the drop stemmed from a single piece of legislation, the Clean Air Act, which required automobiles to switch from leaded gasoline to unleaded. With less lead in the air, crime saw a significant drop twenty-three years later. It turns out high lead levels in the air impair infant brain development, leading to more impulsive behavior and less long-term thinking. Is the correlation between lead levels and crime a coincidence? Likely not. Different countries switched over to unleaded gasoline at different times, and all of them saw the crime rate drop about twenty-three years after making the switch—just when the children raised with less lead were becoming adults.[2] If the hypothesis is correct, it means the Clean Air Act might have done more to fight crime than any other policy in American history. While this hypothesis requires more research, it highlights the importance of the idea that our livewiring process can be insidiously influenced by molecules, hormones, and toxins. If you've ever doubted the significance of brain plasticity, rest assured that its tendrils reach from the individual to the society.

Because of livewiring, we are each a vessel of space and time. We drop into a particular spot on the world and vacuum in the details of that spot. We become, in essence, a recording device for our moment in the world.

When you meet an older person and feel shocked by the opinions or worldview she holds, you can try to empathize with her as a recording device for her window of time and her set of experiences. Someday your brain will be that time-ossified snapshot that frustrates the next generation.

Here's a nugget from my vessel: I remember a song produced in 1985 called "We Are the World." Dozens of superstar musicians performed it to raise money for impoverished children in Africa. The theme was that each of us shares responsibility for the well-being of everyone.

Looking back on the song now, I can't help but see another interpretation through my lens as a neuroscientist. We generally go through life thinking there's *me* and there's *the world*. But as we've seen in this book, who you are emerges from everything you've interacted with: your environment, all of your experiences, your friends, your enemies, your culture, your belief system, your era—all of it. Although we value statements such as "he's his own man" or "she's an independent thinker," there is in fact no way to separate yourself from the rich context in which you're embedded. There is no *you* without the external. Your beliefs and dogmas and aspirations are shaped by it, inside and out, like a sculpture from a block of marble. Thanks to livewiring, each of us is the world.

ACKNOWLEDGMENTS

My career in neuroscience has been enhanced by many people who have mirrored my fascination with the endlessly creative toolbox of nature, and from whom I have learned the thrill of chasing down answers. These are my parents, Cirel and Arthur, who shaped my brain during its most sensitive periods; Read Montague, Terry Sejnowski, and Francis Crick, who further molded it in graduate school and during my postdoctoral fellowship; and scores of friends, students, and colleagues. I thank my colleagues at Stanford for providing a castle of intellectual banquets. There are so many friends to whom I turn for inspiration and good discussion—far too many to list here—but this list includes Don Vaughn, Jonathan Downar, Brett Mensh, and all the students in my lab over the years. I thank Tristan Renz and Scott Freeman for financially supporting our sensory-substitution work before anyone else was willing to take a chance on it. I thank my former graduate student and current business partner Scott Novich for working with me to make the Neosensory technology a reality. And I have constant gratitude for the constantly growing team of Neosensory employees.

I thank Dan Frank and Jamie Byng for being such magnificent publishers and unwavering supporters. I thank the Wylie Agency—especially Andrew, Sarah, James, and Kristina—for strong and ever-supportive backing.

I am grateful to have received close readings of this book from many people, including Mike Perrotta, Shahid Mallick, Sean Judge, and all the wonderful students who take my Brain Plasticity course at Stanford.

I dedicate this book to my two young children, Aristotle and Aviva, in whose cute little heads the principles of brain plasticity play out second by second. And I express my deepest love and gratitude to my wife, Sarah, for being my support and bedrock and reinforcement. Although adoration usually lives in the domain of lyric poetry, it would be nothing without livewiring: our shared love has rewritten each other's brains.

Finally, I want to acknowledge how much my students and my readers around the world have inspired me. They don't always realize how terrific my job is, especially when I have the enviable job of unwrapping an idea they had never thought about before. Reflections off a beautiful truth light up both our faces.

NOTES

To make the ideas in this book widely accessible, I've written the concepts in straightforward language instead of in the argot of the field. This choice has pros and cons. To minimize the cons, the endnotes below allow interested readers to map the concepts to the original literature, deeper details, and scientific vocabulary.

1 The Electric Living Fabric

1. Personal interview with Matthew's family.
2. Strange but true: Matthew's surgeon was Dr. Ben Carson, who later ran for president of the United States on the Republican ticket and lost to Donald Trump.
3. To make matters more complex, the neurons are supported by an equal number of cells called glia. While glia are important for long-term function, neurons are the ones zipping information around rapidly. It used to be thought there were ten times the number of glia as neurons; we now know from new methods (for example, the isotropic fractionator) that the numbers are about one to one. See Von Bartheld CS, Bahney J, Herculano-Houzel S (2016), The search for true numbers of neurons and glial cells in the human brain: A review of 150 years of cell counting, *J Comp Neurol* 524(18): 3865–95. For a general view of the numbers, see also Gordons, *The Synaptic Organization of the Brain* (New York: Oxford University Press, 2004).
4. Here is a tiny subset of experiences a two-year-old might have in a day, all of which shape his future trajectory in some unknowable way: he listens to a story about a boy with a long tail who swats flies. His mother's friend Josette visits with a silver hot pot of homemade steaming meatballs. Three older boys scream past the house on bicycles. He sees a white cat sleeping on the warm hood of a truck. His mother says to his father, "It's like that time in New Mexico," and they both laugh. His father stands over the sink and eats from a Tupperware container of Brussels sprouts, speaking with his mouth full. The boy lays his cheek on the

coolness of the hardwood floor. He sees a large man in a beaver costume handing out peanuts. And so on. Each of these experiences shapes him in some minute way, and if the experiences were slightly different, he would grow into a slightly different man. These considerations might reasonably concern parents with the responsibility of steering a child in the right direction. But the vastness of the ocean of possible experience renders it impossible to navigate. You cannot know the impact of any one book choice over another, or any one decision or exposure. A life trajectory—even in a single day—is far too complex to predict the impacts. While this does nothing to diminish the duties and concerns of a parent, in the end the unknowability can be a little bit liberating.

5. Nishiyama T (2005), Swords into plowshares: Civilian application of wartime military technology in modern Japan, 1945–1964 (PhD diss., Ohio State University).

6. The main argument of this section was touched upon in Eagleman DM (2011), *Incognito: The Secret Lives of the Brain* (New York: Pantheon).

7. There is debate about how to mark clean definitional borders around the term "plasticity." How long does the change need to last to be labeled plastic? Is it possible to distinguish plasticity from concepts such as maturation, predisposition, flexibility, and elasticity? These semantic debates are tangential to the point of this book; nonetheless, for those interested I have included some discussion here.

   Part of the debate revolves around *when* to use the term "plasticity." Are issues of developmental plasticity, phenotypic plasticity, and synaptic plasticity expressions of the same thing, or is "plasticity" a single term used sloppily in different contexts? To my knowledge, the first person to explicitly address this question was Jacques Paillard, in his 1976 essay "Réflexions sur l'usage du concept de plasticité en neurobiologie," which was translated and commented upon by Bruno Will and colleagues in 2008. Following the lead of Paillard, the 2008 paper suggests that a proper example of "plasticity" must include both structural *and* functional changes (not just one or the other) and should be distinguished from *flexibility* (such as a preprogrammed adaptation), *maturation* (say, the normal unpacking of an organism), and *elasticity* (short-term changes that eventually return to their former state). As we will see later in this book, these themes are not always possible to distinguish. As one example, we'll spend an entire chapter exploring how the brain changes at many different timescales, and how such changes can be passed along to different parts of the system (for example, from the level of the molecular to the larger cellular architecture). In this light, if our technology measured a change that eventually returned to its original state—but only because we couldn't simultaneously measure the ripple effects—would we have to conclude the whole system is merely elastic, and not plastic? It seems to me unwise to anchor our semantic definitions to our current technologies.

   Debates about the word "plasticity" often tend toward tempests in a teapot. In the context of this book, our interest is to understand the self-modification of the three pounds of futuristic technology in our skulls. If you have a rich understanding of that by the end of this book, I will have succeeded.

8. Matthew's limp is on the side opposite the removed hemisphere, because each hemisphere controls the opposite side of the body. The residual limp comes from the fact that the remaining hemisphere was partially able to take over the motor function of the removed hemisphere, but not entirely.

2 Just Add World

1. Gopnik A, Schulz L (2004), Mechanisms of theory formation in young children, *Trends Cogn Sci* 8:371–77.
2. Spurzheim J (1815), *The Physiognomical System of Drs. Gall and Spurzheim*, 2nd ed. (London: Baldwin, Cradock and Joy).
3. Darwin C (1874), *The Descent of Man* (Chicago: Rand, McNally).
4. Bennett EL et al. (1964), Chemical and anatomical plasticity of brain, *Science* 164:610–19.
5. Diamond M (1988), *Enriching Heredity* (New York: Free Press).
6. Rosenzweig MR, Bennett EL (1996), Psychobiology of plasticity: Effects of training and experience on brain and behavior, *Behav Brain Res* 78:57–65; Diamond M (2001), Response of the brain to enrichment, *An Acad Bras Ciênc* 73:211–20.
7. Jacobs B, Schall M, Scheibel AB (1993), A quantitative dendritic analysis of Wernicke's area in humans. II. Gender, hemispheric, and environmental factors, *J Comp Neurol* 327:97–111. Now, you will wisely ask which way the arrow of causality goes here: Perhaps those with better dendrites were better able to win admission to college, rather than college causing the growth? Good question. We don't have the experiments yet to rule that out. But as we'll see in the coming chapters, brain changes can now be measured on the fly as people learn new things, including juggling, music, navigation, and more.
8. The Human Genome Project originally estimated about twenty-four thousand genes; the number has bounced around since then, going as low as nineteen thousand. See Ezkurdia I et al. (2014), Multiple evidence strands suggest that there may be as few as 19,000 human protein-coding genes, *Hum Mol Genet* 23(22): 5866–78.
9. Much more on this in later chapters. While experience-dependence and experience-independence seem like opposite stories, there is not always a sharp border between them (see Cline H [2003], Sperry and Hebb: Oil and vinegar?, *Trends Neurosci* 26[12]: 655–61). Sometimes hardwired mechanisms will simulate world experience, and other times world experience leads to gene expression that leads to new hardwiring. Consider what appears to be a clear story of activity-dependent activity: in the primary visual cortex one finds alternating bands of tissue that carry information from the left eye and the right eye (more on this in later chapters). Axons carrying this eye-specific visual information initially branch widely in the cortex and then segregate into eye-specific patches. How do they know how to segregate? The trick is that the separation emerges from pat-

terns of correlated activity: neurons in the left eye tend to be more correlated with each other than they are with neurons in the right eye.

In the mid-1960s, the Harvard neurobiologists David Hubel and Torsten Wiesel showed that the map of evenly alternating stripes could be drastically changed by experience: shutting one eye of an animal leads to an expansion of the territory occupied by fibers from the open eye, demonstrating the need for neural activity in the synaptic competition that forms these maps (Hubel DH, Wiesel TN [1965], Binocular interaction in striate cortex of kittens reared with artificial squint, *J Neurophysiol* 28:1041–59).

However, there was a mystery buried in here, because Hubel and Wiesel had previously observed that the establishment of the alternating territories of left and right eyes did not depend on activity: even animals raised in total darkness developed these patterns (Horton JC, Hocking DR [1996]. An adult-like pattern of ocular dominance columns in striate cortex of newborn monkeys prior to visual experience, *J Neurosci.* 16[5]:1791–807). How could these findings be consistent?

It took years before the paradox was resolved. It turned out that while the developing animal floats in the womb, its retina generates spontaneous waves of activity. These waves roughly simulate vision. The swells of activity are crude—they do not have the sharp borders of real, fine-grained visual experience—but they are sufficient to correlate the activity in neighboring fibers from each eye, causing eye-specific segregation in the later brain areas (such as the lateral geniculate nucleus of the thalamus, and the cortex). In other words, the brain supplies its own activity early in development to help in the process of segregating the eyes; later, visual input from the outside world takes over (Meister M et al. [1991], Synchronous bursts of action potentials in ganglion cells of the developing mammalian retina, *Science* 252(5008): 939–43). Thus, there is a blur between world experience and prespecified neuronal activity. The interplay between world experience and genetic instruction can be complex. The overall principle is that experience-independent molecular mechanisms lead to an initial, imprecise wiring of the brain. Later, activity caused by interaction with the world refines those connections. We can no longer think about the brain as the result of only genes or only world experience, because sometimes the genes impersonate world experience. The experience-dependent and experience-independent mechanisms are tightly intertwined.

10. Leonhard K (1970), Kaspar Hauser und die moderne Kenntnis des Hospitalismus, *Confin Psychiat* 13:213–29.

11. DeGregory L (2008), The girl in the window, *St. Petersburg Times*. It should be noted that in recent years Danielle has shown some improvements. She has learned how to use the toilet, can understand some of what people say to her, and can make some limited verbal replies. Recently, she was attending prekindergarten and learning to trace letters. These are lovely and welcome signs; unfortunately, it remains unlikely that she'll be able to gain much of the ground lost during her tragic first years of life.

One more thing to note. Children like Danielle, who are raised in conditions of stress and deprivation, do not grow properly in terms of their bodies; this is known as psychosocial dwarfism. Ironically, one medical writer in the 1990s tried to coin a new term for this: "Kaspar Hauser syndrome" (Money J (1992). *The Kaspar Hauser syndrome of "psychosocial dwarfism": Deficient statural, intellectual, and social growth induced by child abuse.* Prometheus Books. 1992), an unfortunate choice, given that Hauser was almost certainly faking a feral past.

12. Fortunately, current animal rights protocols prohibit research like this from happening today. Even at the time, many of Harlow's colleagues were horrified by his experiments, which invigorated the burgeoning animal liberation movement in the United States. One of Harlow's critics, Wayne Booth, wrote that Harlow's experiments merely proved "what we all knew in advance—that social creatures can be destroyed by destroying their social ties."

3 The Inside Mirrors the Outside

1. See Penfield W (1952), Memory mechanisms, *AMA Arch Neurol Psychiatry* 67(2): 178–98; Penfield W (1961), Activation of the record of human experience, *Ann R Coll Surg Engl* 29(2): 77–84.

2. The cortex is the outer layer—usually about three millimeters thick—of the brain. It is referred to as gray matter because its cells look darker than the white matter beneath the cortex. In larger animals the cortex is usually folded and grooved. The first strip where Penfield measured is called the somatosensory cortex, referring to sensation from the body, or soma.

3. Ettlin D (1981), Taub denies allegations of cruelty, *Baltimore Sun*, Nov. 1, 1981.

4. Pons TP et al. (1991), Massive cortical reorganization after sensory deafferentation in adult macaques, *Science* 252:1857–60; Merzenich M (1998), Long-term change of mind, *Science* 282(5391): 1062–63; Jones EG, Pons TP (1998), Thalamic and brainstem contributions to large-scale plasticity of primate somatosensory cortex, *Science* 282(5391): 1121–5; Merzenich M et al. (1984), Somatosensory cortical map changes following digit amputation in adult monkeys, *J Comp Neurol* 224:591–605.

5. Beyond the cortex, there were massive reorganizational shifts in other brain areas as well, such as the thalamus and the brainstem; we'll return to that later.

6. Knight R (2005), *The Pursuit of Victory: The Life and Achievement of Horatio Nelson* (New York: Basic Books).

7. Mitchell SW (1872), *Injuries of Nerves and Their Consequences* (Philadelphia: Lippincott).

8. Such techniques started off with magnetoencephalography (MEG) and soon moved to functional magnetic resonance imaging, or fMRI. For a review of brain-imaging techniques, see Eagleman DM, Downar J (2015), *Brain and Behavior* (New York: Oxford University Press).

9. Phantom pain teaches us that although brains rewrite their maps, the changes are imperfect: although neurons that previously encoded the arm now encode the face, other neurons downstream still think they are getting arm information. As a result of the downstream confusion, amputees typically feel pain in their phantom limb. Generally, larger cortical changes translate into greater experienced pain. See Flor et al. (1995), Phantom-limb pain as a perceptual correlate of cortical reorganization following arm amputation, *Nature* 375(6531): 482–84; Karl A et al. (2001), Reorganization of motor and somatosensory cortex in upper extremity amputees with phantom limb pain, *J Neurosci* 21:3609–18. We'll understand more about phantom pain later, when we see how different brain areas change at different speeds.

10. Singh AK et al. (2018), Why does the cortex reorganize after sensory loss?, *Trends Cogn Sci* 22(7): 569–82; Ramachandran VS et al. (1992), Perceptual correlates of massive cortical reorganization, *Science* 258:1159–60; Barinaga M (1992), The brain remaps its own contours, *Science* 258:216–18; Borsook D et al. (1998), Acute plasticity in the human somatosensory cortex following amputation, *Neuroreport* 9:1013–17.

11. Weiss T et al. (2004), Rapid functional plasticity in the primary somatomotor cortex and perceptual changes after nerve block, *Eur J Neurosci* 20:3413–23.

12. Clark SA et al. (1988), Receptive-fields in the body-surface map in adult cortex defined by temporally correlated inputs, *Nature* 332:444–45.

13. This rule, known as Hebb's rule, was first proposed in 1949. Hebb DO (1949), *The Organization of Behavior* (New York: Wiley & Sons). It often turns out to be slightly more complex: if neuron A fires just before neuron B, then the bond between them is strengthened; if A fires just after B, their bond is weakened. This is known as spike-timing-dependent plasticity.

14. There are also genetic tendencies that cause the map to form in certain ways; for example, the reason the head is on one end of the map and the feet on another has to do with the way the fibers attach from the body. But this book emphasizes the surprising ways that experience changes the wiring.

15. For historical exactitude: The Louisiana Territory first went to the Spanish. Then in 1802 Spain returned Louisiana to France. But Napoleon sold it to the United States in 1803, because at this point he'd given up the dream of the New World.

16. Elbert T, Rockstroh B (2004), Reorganization of human cerebral cortex: The range of changes following use and injury, *Neuroscientist* 10:129–41; Pascual-Leone A et al. (2005), The plastic human brain cortex, *Annu Rev Neurosci* 28:377–401; D'Angiulli A and Waraich P (2002), Enhanced tactile encoding and memory recognition in congenital blindness, *Int J Rehabil Res* 25(2): 143–45; Collignon O et al. (2006), Improved selective and divided spatial attention in early blind subjects, *Brain Res* 1075(1): 175–82; Collignon O et al. (2009), Cross-modal plasticity for the spatial processing of sounds in visually deprived subjects, *Exp Brain Res* 192(3): 343–58; Bubic A, Striem-Amit E, Amedi A (2010), Large-scale brain plasticity following blindness and the use of sensory substitution devices, in *Multisensory*

Object Perception in the Primate Brain, ed. MJ Naumer and J Kaiser (New York: Springer), 351–80.

17. Amedi A et al. (2010), Cortical activity during tactile exploration of objects in blind and sighted humans, *Restor Neurol Neurosci* 28(2): 143–56; Sathian K, Stilla R (2010), Cross-modal plasticity of tactile perception in blindness, *Restor Neurol Neurosci* 28(2): 271–81. Note also that these changes can be read out in other ways: for example, in a blind Braille reader, a pulse of magnetic stimulation over the occipital cortex will induce a tactile sensation in the fingers (while the pulse has no such effect in sighted control subjects). See Ptito M et al. (2008), TMS of the occipital cortex induces tactile sensations in the fingers of blind Braille readers, *Exp Brain Res* 184(2): 193–200.

18. Hamilton R et al. (2000), Alexia for Braille following bilateral occipital stroke in an early blind woman, *Neuroreport* 11(2): 237–40.

19. Voss P et al. (2006), A positron emission tomography study during auditory localization by late-onset blind individuals, *Neuroreport* 17(4): 383–88; Voss P et al. (2008), Differential occipital responses in early- and late-blind individuals during a sound-source discrimination task, *Neuroimage* 40(2): 746–58. In a second experiment in the same paper, participants guessed at the location of a sound, and the same thing was found: activation of the visual cortex.

20. Renier L, De Volder AG, Rauschecker JP (2014), Cortical plasticity and preserved function in early blindness, *Neurosci Biobehav Rev* 41:53–63; Raz N, Amedi A, Zohary E (2005), V1 activation in congenitally blind humans is associated with episodic retrieval, *Cereb Cortex* 15:1459–68; Merabet LB, Pascual-Leone A (2010), Neural reorganization following sensory loss: The opportunity of change, *Nat Rev Neurosci* 11(1): 44–52.

    As a side note, the connection can be demonstrated in the other direction: when activity in the occipital lobe of blind people is temporarily disrupted (by stimulation with a magnetic pulse), their Braille reading, and even their verbal processing, suffer for it. See Amedi A et al. (2004), Transcranial magnetic stimulation of the occipital pole interferes with verbal processing in blind subjects, *Nat Neurosci* 7:1266.

21. This area is known as the VWFA (visual word form area). Reich L et al. (2011), A ventral visual stream reading center independent of visual experience, *Curr Biol* 21:363–68; Striem-Amit E et al. (2012), Reading with sounds: Sensory substitution selectively activates the visual word form area in the blind, *Neuron* 76:640–52.

22. This area is known as MT (middle temporal) or V5. Ptito M et al. (2009), Recruitment of the middle temporal area by tactile motion in congenital blindness, *Neuroreport* 20:543–47. Matteau I et al. (2010), Beyond visual, aural, and haptic movement perception: hMT+ is activated by electrotactile motion stimulation of the tongue in sighted and in congenitally blind individuals, *Brain Res Bull* 82:264–70.

23. This area is known as LOC (lateral occipital cortex). Amedi et al. (2010).

24. Another way of phrasing this: the brain is a "metamodal" operator. Metamodal means the operations are independent of the specific modes (or senses) that get the information in there. See Pascual-Leone A, Hamilton R (2001), The metamodal organization of the brain, *Prog Brain Res* 134:427–45; Reich L, Maidenbaum S, Amedi A (2011), The brain as a flexible task machine: Implications for visual rehabilitation using noninvasive vs. invasive approaches, *Curr Opin Neurol* 25:86–95. Also see Maidenbaum S et al. (2014), Sensory substitution: Closing the gap between basic research and widespread practical visual rehabilitation, *Neurosci Biobehav Rev* 41:3–15; Reich L et al. (2011), A ventral visual stream reading center independent of visual experience, *Curr Biol* 21(5): 363–8; Striem-Amit E et al. (2012), The large-scale organization of "visual" streams emerges without visual experience, *Cereb Cortex* 22(7): 1698–709; Meredith MA et al. (2011), Crossmodal reorganization in the early deaf switches sensory, but not behavioral roles of auditory cortex, *Proc Natl Acad Sci USA* 108(21): 8856–61; Bola Ł et al. (2017), Task-specific reorganization of the auditory cortex in deaf humans, *Proc Natl Acad Sci USA* 114(4): E600–E609. For reviews, see Bavelier and Hirshorn (2010) and Dormal, Collignon (2011).

25. Finney EM, Fine I, Dobkins KR (2001), Visual stimuli activate auditory cortex in the deaf, *Nat Neurosci* 4(12): 1171–73; Meredith MA et al. (2011).

26. Elbert, Rockstroh (2004); Pascual-Leone et al. (2005).

27. See Hamilton RH, Pascual-Leone A, Schlaug G (2004), Absolute pitch in blind musicians, *Neuroreport* 15:803–6; Gougoux F et al. (2004), Neuropsychology: Pitch discrimination in the early blind, *Nature* 430(6997): 309.

28. Voss et al. (2008).

29. Ben died in 2016, at the age of sixteen, when the cancer that had taken both his eyes returned.

30. *Extraordinary People,*: The Boy Who Sees Without Eyes" Season 1, episode 43, aired on Jan. 29, 2007.

31. Teng S, Puri A, Whitney D (2012), Ultrafine spatial acuity of blind expert human echolocators, *Exp Brain Res* 216(4): 483–88; Schenkman BN, Nilsson ME (2010), Human echolocation: Blind and sighted persons' ability to detect sounds recorded in the presence of a reflecting object, *Perception* 39(4): 483; Arnott SR et al. (2013), Shape-specific activation of occipital cortex in an early blind echolocation expert, *Neuropsychologia* 51(5): 938–49; Thaler L et al. (2014), Neural correlates of motion processing through echolocation, source hearing, and vision in blind echolocation experts and sighted echolocation novices, *J Neurophysiol* 111(1): 112–27. Also, in blind echolocators, listening to sound echoes activates the visual rather than the auditory cortex: Thaler L et al. (2011), Neural correlates of natural human echolocation in early and late blind echolocation experts, *PLoS One* 6(5): e20162. Echolocation can be improved by technology: several new projects employ an ultrasonic sensor mounted onto a wearable pair of glasses that measures the distance to the nearest object and relays it into a clear audio signal that plays different tones to represent different distances.

32. Griffin DR (1944), Echolocation by blind men, bats, and radar, *Science* 100(2609): 589–90.

33. Amedi A et al. (2003), Early "visual" cortex activation correlates with superior verbal-memory performance in the blind, *Nat Neurosci* 6:758–66.

34. In other words, the task of grayscale discrimination takes over a plot of cortex that is usually devoted to gray *and* color.

35. Kok MA et al. (2014), Cross-modal reorganization of cortical afferents to dorsal auditory cortex following early- and late-onset deafness, *J Comp Neurol* 522(3): 654–75; Finney EM et al. (2001), Visual stimuli activate auditory cortex in the deaf, *Nat Neurosci* 4(12): 1171.

36. In autism, regions of the brain grow at different rates, apparently causing regions to establish abnormal connectivity—with the end result that the long-distance connections in the autistic brain are subtly different, leading to deficits in language and social behavior. Redcay E, Courchesne E (2005), When is the brain enlarged in autism? A meta-analysis of all brain size reports, *Biol Psychiatry* 58:1–9. In other words, a livewired system can unpack itself from a single cell, but the *way* it unpacks—the exact rhythm and order—results in different outcomes. We should note that theories on autism are wide-ranging, including defects in the mirror neuron system, vaccines, underconnectivity, disorders of central coherence, and many more. So that idea that it is simply a redistribution of cortex is likely to be only a part of the story. Nonetheless, see examples such as Boddaert N et al. (2005), Autism: Functional brain mapping of exceptional calendar capacity, *Br J Psychiatry* 187:83–86; LeBlanc j, Fagiolini M (2011), Autism: A "critical period" disorder?, *Neural plasticity.* 2011:921680.

37. Voss et al. (2008).

38. Pascual-Leone A, Hamilton R (2001), The metamodal organization of the brain, in *Vision: From Neurons to Cognition*, ed. C Casanova and M Ptito (New York: Elsevier Science), 427–45.

39. Merabet LB et al. (2008), Rapid and reversible recruitment of early visual cortex for touch, *PLoS One* 3(8): e3046. Also note that an early version of these results was published in Pascual-Leone and Hamilton (2001).

40. Merabet LB et al. (2007), Combined activation and deactivation of visual cortex during tactile sensory processing, *J Neurophysiol* 97:1633–41.

41. Although some forms of dreaming can occur during non-REM sleep (Kleitman N [1963], *Sleep and Wakefulness.* [Chicago: U Chicago Press], such dreams are quite different from the more common REM dreams: they are usually related to plans or elaborate thoughts, and they lack the visual vividness and hallucinatory and delusory components of REM dreams. Because our proposal depends on the strong activation of the visual system, REM sleep is implicated over non-REM sleep.

42. This activity is called PGO waves (ponto-geniculo-occipital waves), so named because they originate in a brain area called the pons, then travel to the lateral geniculate nucleus (hence the "geniculo"), and then complete their journey to

the occipital (visual) cortex. As a side note, there is some debate (but not much) about whether PGO waves, REM sleep, and dreaming are actually equivalent, or whether they are separable issues. For completeness here, I mention that children and schizophrenics with prefrontal lobotomies can have REM sleep with very little dreaming. See Solms M (2000), Dreaming and REM sleep are controlled by different brain mechanisms, *Behav Brain Sci* 23(6): 843–50 (including the spirited discussion by colleagues that accompanies the paper); also see Jus et al. (1973), Studies on dream recall in chronic schizophrenic patients after prefrontal lobotomy, *Biol Psychiatry* 6(3): 275–93. Also, it isn't known whether the brainstem activity is random, or reflects the day's memories, or serves to practice neural programs—but here the important part is that once the waves propagate to the visual areas, the activity is experienced as visual. See Nir Y, Tononi G (2010), Dreaming and the brain: From phenomenology to neurophysiology, *Trends Cogn Sci* 14(2): 88–100.

43. Eagleman DM, Vaughn DA (2020, under review). Like any biological theory, this one must be understood in the context of evolutionary time vistas. The programs to build this wiring are deep in the genetics, and therefore not dependent on the experiences of an individual lifetime. Because these circuits evolved over hundreds of millions of years, they are unaffected by our modern ability to defy darkness with electrical light.

Our hypothesis leaves open many questions: For example, why don't we dream constantly, but instead in bursts? It also does not reflect on the mystery of dream *content*. For one overview of the content topic, see Flanagan O (2000), *Dreaming Souls: Sleep, Dreams, and the Evolution of the Conscious Mind* (New York: Oxford University Press). Our hypothesis will be further studied in the future by examining the visual changes in diseases that lead to the loss or impairment of dreaming. There is much more to uncover here, given the opportunity to understand dreaming through a new lens. Also, REM sleep can be suppressed by monoamine oxidase inhibitors, or by certain brain lesions, and yet it is difficult to detect any problems (cognitive or physiological) in people with REM sleep problems. (Siegel JM [2001], The REM sleep-memory consolidation hypothesis, *Science* 294:1058–63). However, our hypothesis predicts *visual* problems, and in fact this is precisely what's observed in people who go on monoamine oxidase inhibitors or tricyclic antidepressants. Some physicians suggest the blurred vision results from dry eyes; we suggest that may not be the correct root of the problem.

Another interesting technical point: various hypotheses in the past have suggested that the duration of REM sleep has something to do with the previous waking period. But if that were the case, one might expect the duration of REM sleep to be longer at the beginning of the night and shorter later. What actually happens is just the opposite. Your first REM sleep period of the night may last only five to ten minutes, while your last one can be more than twenty-five minutes. (See Siegel JM [2005], Clues to the functions of mammalian sleep, *Nature*

437[7063]:1264–71). This is consistent with a system that has to fight harder the longer it's been without visual input.

One more point: In a young animal, if you reduce the light entering one eye (but not the other), you can measure a takeover of territory from the good eye. If you then deprive the animal of REM sleep during the critical period of susceptibility, the imbalance is accelerated. In other words, REM sleep (which benefits both visual channels equally) helps to slow takeovers—in this case the takeover of one eye's territory over another's. Without the REM sleep, the takeover happens more quickly.

44. In a 1999 paper titled "The Dreams of Blind Men and Women," Craig Hurovitz and colleagues carefully recorded and analyzed the details of 372 dreams from fifteen blind adults.

45. People who become blind *after* the age of seven have more visual content in their dreams than those who become blind earlier: Amadeo M, Gomez E (1966), Eye movements, attention and dreaming in the congenitally blind, *Can Psychiat Assoc J.*: 501–7; Berger RJ et al. (1962), The eec, eye-movements and dreams of the blind, *Quart J Exp Psychol* 14(3): 183–6; Kerr NH et al. (1982), The structure of laboratory dream reports in blind and sighted subjects, *J Nerv Mental Dis* 170(5): 286–94; Hurovitz C et al. (1999), The dreams of blind men and women: A replication and extension of previous findings, *Dreaming* 9:183–93; Kirtley DD (1975), *The Psychology of Blindness*. (Chicago: Nelson-Hall). The occipital lobe in the late blind is less fully conquered by other senses: see, for example, Voss et al. (2006, 2008).

46. Zepelin H, Siegel JM, Tobler I (2005), in *Principles and Practice of Sleep Medicine*, vol. 4, ed. MH Kryger, T Roth, and WC Dement (Philadelphia: Elsevier Saunders), 91–100. Jouvet-Mounier D, Astic L, Lacote D (1970), Ontogenesis of the states of sleep in rat, cat, and guinea pig during the first postnatal month, *Dev Psychobiol* 2:216–39.

47. Siegel JM (2005).

48. Angerhausen D et al. (2012), An astrobiological experiment to explore the habitability of tidally locked m-dwarf planets, *Proc Int Astron Union* 8(S293): 192–96. Note that this would look something like the way the same face of our moon always faces the earth. It should be noted, however, that our moon's rotation time is exactly the same as its orbit time, which is why we always see the same face, but the moon does still have days and nights, because it faces the sun differently. A planet tidally locked to a star has no days or nights.

4 Wrapping Around the Inputs

1. See Chorost M (2005), *Rebuilt: How Becoming Part Computer Made Me More Human* (Boston: Houghton Mifflin); Chorost M (2011), *World Wide Mind: The*

Coming Integration of Humanity, Machines, and the Internet (New York: Free Press). See also Chorost M (2005), My bionic quest for Bolero, *Wired*.

2. Fleming N (2007), How one man "saw" his son after 13 years, *Telegraph*.

3. Ahuja AK et al. (2011), Blind subjects implanted with the Argus II retinal prosthesis are able to improve performance in a spatial-motor task, *Br J Ophthalmol* 95(4): 539–43.

4. My analogy limps slightly, because plug and play in the computer world is accomplished by having agreed-upon rules of engagement: the peripheral comes with some information about itself, and it tells that to the computer so that the central processor knows what to do. In contrast, the brain uses a somewhat different protocol. Presumably, peripheral devices like the eyes know nothing about themselves. They simply do what they do. But the brain has the capacity to learn how to extract useful information from them—in other words, how to *use* them.

5. Photograph by Sharon Steinmann, AL.com. Alabama baby born without a nose, mom says he's perfect, ABC News, www.abcnews.go.com.

6. Lourgos AL (2015), Family of Peoria baby born without eyes prepares for treatment in Chicago, *Chicago Tribune*, www.chicagotribune.com.

7. This is called LAMM syndrome; LAMM stands for labyrinthine aplasia, microtia, and microdontia. It affects development of the ears and teeth. LAMM syndrome is also characterized by small outer ears and small, gapped teeth—because the gene that is mutated (FGF3) triggers a cascade of cellular reactions that lead to the formation of the structures of the inner ear, the outer ear, and the teeth. When FGF3 is mutated, it doesn't give the proper go signal, and the ears and teeth of LAMM syndrome result.

8. Wetzel F (2013), Woman born without tongue has op so she can speak, eat, and breathe more easily, *Sun*, Jan. 18, 2013.

9. This is generally known as congenital insensitivity to pain or congenital analgesia. See Eagleman DM, Downar J (2015), *Brain and Behavior* (New York: Oxford University Press).

10. Abrams M, Winters D (2003), Can you see with your tongue?, *Discover*.

11. Macpherson F, ed. (2018), *Sensory Substitution and Augmentation* (Oxford: Oxford University Press); Lenay C et al. (2003), Sensory substitution: Limits and perspectives, in *Touching for Knowing: Cognitive Psychology of Haptic Manual Perception*, ed. Y Hatwell, A Streri, and E Gentaz (Philadelphia: John Benjamins), 275–92; Poirier C, De Volder AG, Scheiber C (2007), What neuroimaging tells us about sensory substitution, *Neurosci Biobehav Rev* 31:1064–70; Bubic A, Striem-Amit E, Amedi A (2010), Large-scale brain plasticity following blindness and the use of sensory substitution devices, in *Multisensory Object Perception in the Primate Brain*, ed. MJ Naumer and J Kaiser (New York: Springer), 351–80; Novich SD, Eagleman DM (2015), Using space and time to encode vibrotactile information: Toward an estimate of the skin's achievable throughput, *Exp Brain Res* 233(10): 2777–88; Chebat DR et al. (2018), Sensory substitution and the neural correlates of

navigation in blindness, in *Mobility of Visually Impaired People* (Cham: Springer), 167–200.

12. Bach-y-Rita P (1972), *Brain Mechanisms in Sensory Substitution* (New York: Academic Press); Brain mechanisms in sensory substitution. Bach-y-Rita P (2004), Tactile sensory substitution studies, *Ann NY Acad Sci* 1013:83–91.

13. Hurley S, Noë A (2003), Neural plasticity and consciousness, *Biology and Philosophy* 18(1): 131–68; Noë A (2004), *Action in Perception* (Cambridge, Mass: MIT Press).

14. Bach-y-Rita P et al. (2003), Seeing with the brain, *Int J Human-Computer Interaction*, 15(2): 285–95; Nagel SK et al. (2005), Beyond sensory substitution—learning the sixth sense, *J Neural Eng* 2(4): R13–R26.

15. Starkiewicz W, Kuliszewski T (1963), The 80-channel elektroftalm, in *Proceedings of the International Congress on Technology and Blindness* (New York: American Foundation for the Blind).

16. This idea of the cortex being fundamentally the same everywhere—but being shaped by its inputs—was originally explored by the neurophysiologist Vernon Mountcastle, and later reinvigorated by the scientist and inventor Jeff Hawkins. See Hawkins J, Blakeslee S (2005), *On Intelligence* (New York: Times Books).

17. Pascual-Leone A, Hamilton R (2001), The metamodal organization of the brain, in *Vision: From Neurons to Cognition,* ed. C Casanova and M Ptito (New York: Elsevier Science), 427–45.

18. Sur M (2001), Cortical development: Transplantation and rewiring studies, in *International Encyclopedia of the Social and Behavioral Sciences,* ed. N Smelser and P Baltes (New York: Elsevier).

19. Sharma J, Angelucci A, Sur M (2000), Induction of visual orientation modules in auditory cortex, *Nature* 404:841–47. The cells in the new auditory cortex now responded, for example, to different orientations of lines.

20. There's a caveat here, one that we'll unpack more in later chapters: the brain does not arrive as a totally blank slate. And this is why the visually responsive auditory cortex in the ferret came out somewhat sloppier in its coding than the traditional visual cortex. Local genetics make certain areas slightly more predisposed to certain types of sensory input. There is a continuum between firm building plans (genetics) and flexibility from activity (livewiring). Why? Because on evolutionary timescales, stable inputs slowly move from something learned in a lifetime to genetically preprogrammed. Our goal here is to concentrate on the tremendous flexibility seen within a lifetime.

21. Bach-y-Rita P et al. (2005), Late human brain plasticity: Vestibular substitution with a tongue BrainPort human-machine interface, *Intellectica* 1(40): 115–22; Nau AC et al. (2015), Acquisition of visual perception in blind adults using the Brain-Port artificial vision device, *Am J Occup Ther* 69(1): 1–8; Stronks HC et al. (2016), Visual task performance in the blind with the BrainPort V100 Vision Aid, *Expert Rev Med Devices* 13(10): 919–31.

22. Sampaio E, Maris S, Bach-y-Rita P (2001), Brain plasticity: "Visual" acuity of blind persons via the tongue, *Brain Res* 908(2): 204–7.

23. Levy B (2008), The blind climber who "sees" with his tongue, *Discover,* June 22, 2008.

24. Bach-y-Rita P et al. (1969), Vision substitution by tactile image projection, *Nature* 221:963–64; Bach-y-Rita P (2004), Tactile sensory substitution studies, *Ann NY Acad Sci* 1013:83–91.

25. This is an area called MT+. Matteau I et al. (2010), Beyond visual, aural, and haptic movement perception: hMT+ is activated by electrotactile motion stimulation of the tongue in sighted and in congenitally blind individuals, *Brain Res Bull* 82(5–6): 264–70. See also Amedi A et al. (2010), Cortical activity during tactile exploration of objects in blind and sighted humans, *Restor Neurol Neurosci* 28(2): 143–56; and Merabet L et al. (2009), Functional recruitment of visual cortex for sound encoded object identification in the blind, *Neuroreport* 20(2): 132. In the blind, many other areas in the occipital cortex become active as well, just as we would expect from the cortical real estate takeovers we saw in the previous chapter.

26. WIRED Science video: "Mixed Feelings."

27. The Forehead Retina System was developed by EyePlusPlus Inc. of Japan and Tachi Laboratory at the University of Tokyo. It uses edge enhancement and temporal band-pass filtering to impersonate the retina.

28. This is a good way to prevent the waist from going to waste. Lobo L et al. (2018), Sensory substitution: Using a vibrotactile device to orient and walk to targets, *J Exp Psychol Appl* 24(1): 108. Also see Lobo L et al. (2017), Sensory substitution and walking toward targets: An experiment with blind participants. Their research demonstrates that the blind participants' walking trajectories are not preplanned, but instead emerge dynamically as new information streams in.

29. See Kay L (2000), Auditory perception of objects by blind persons, using a bio-acoustic high resolution air sonar, *J Acoust Soc Am* 107(6): 3266–76. The sonic glasses debuted in the mid-1970s and were improved by many steps since then (see Kay's Binaural Sensory Aid and the later KASPA system, which represents surface texture by timbre). The resolution of ultrasound techniques is not high, especially in the up-down direction—making the sonic glasses mostly useful for detecting objects in a narrow horizontal slot.

30. Bower TGR (1978), Perceptual development: Object and space, in *Handbook of Perception,* vol. 8, *Perceptual Coding,* ed. EC Carterette and MP Friedman (New York: Academic Press). See also Aitken S, Bower TGR (1982), Intersensory substitution in the blind, *J Exp Child Psychol* 33:309–23.

31. Because of the diminishing plasticity with age, sensory substitution has to be individualized—both for the current age and for the age at which blindness was acquired. Bubic, Striem-Amit, Amedi (2010).

32. Meijer PB (1992), An experimental system for auditory image representations, *IEEE Trans Biomed Eng* 39(2): 112–21.

33. See technical details and hear demonstrations of the vOICe algorithm at www .seeingwithsound.com.

34. Arno P et al. (1999), Auditory coding of visual patterns for the blind, *Perception* 28(8): 1013–29; Arno P et al. (2001), Occipital activation by pattern recognition in the early blind using auditory substitution for vision, *Neuroimage* 13(4): 632–45; Auvray M, Hanneton S, O'Regan JK (2007), Learning to perceive with a visuo-auditory substitution system: Localisation and object recognition with "the vOICe," *Perception* 36:416–30; Proulx MJ et al. (2008), Seeing "where" through the ears: Effects of learning-by-doing and long-term sensory deprivation on localization based on image-to-sound substitution, *PLoS One* 3(3): e1840.

35. Cronly-Dillon J, Persaud K, Gregory RP (1999), The perception of visual images encoded in musical form: A study in cross-modality information transfer, *Proc Biol Sci* 266(1436): 2427–33; Cronly-Dillon J, Persaud KC, Blore R (2000), Blind subjects construct conscious mental images of visual scenes encoded in musical form, *Proc Biol Sci* 267(1458): 2231–38.

36. Quotation from Pat Fletcher in an article in ACB's Braille forum, as cited in Maidenbaum S et al. (2014), Sensory substitution: Closing the gap between basic research and widespread practical visual rehabilitation, *Neurosci Biobehav Rev* 41:3–15.

37. Specifically, Amedi et al. (2007) demonstrated activation in a brain region known as the lateral-occipital tactile-visual area (LOtv). This region appears to encode information about shape—whether activated by vision, touch, or the learning of a visual-to-auditory soundscape. Amedi A et al. (2007), Shape conveyed by visual-to-auditory sensory substitution activates the lateral occipital complex, *Nat Neurosci* 10:687–89. See a summary of the experience of one user in Piore A (2017), *The Body Builders: Inside the Science of the Engineered Human* (New York: Ecco).

38. Collignon O et al. (2007), Functional cerebral reorganization for auditory spatial processing and auditory substitution of vision in early blind subjects, *Cereb Cortex* 17(2): 457–65.

39. Abboud S et al. (2014), EyeMusic: Introducing a "visual" colorful experience for the blind using auditory sensory substitution, *Restor Neurol Neurosci* 32(2): 247–57. EyeMusic is based on an older technology called SmartSight: Cronly-Dillon et al. (1999, 2000).

40. Massiceti D, Hicks SL, van Rheede JJ (2018), Stereosonic vision: Exploring visual-to-auditory sensory substitution mappings in an immersive virtual reality navigation paradigm, *PLoS One* 13(7): e0199389. Tapu R, Mocanu B, Zaharia T (2018), Wearable assistive devices for visually impaired: A state of the art survey, *Pattern Recognit Lett*; Kubanek M, Bobulski J (2018), Device for acoustic support of orientation in the surroundings for blind people, *Sensors* 18(12): 4309. See also Hoffmann R et al. (2018), Evaluation of an audio-haptic sensory substitution device for enhancing spatial awareness for the visually impaired, *Optom Vis Sci* 95(9): 757.

41. Trachoma, the leading cause of blindness in the developing world, has left almost two million people blind. The second-leading cause of blindness, onchocerciasis,

is endemic in thirty African countries. Many researchers are considering using sensory-substitution software as a bridge to relearning eyesight in conjunction with other therapies (for example, cornea surgery).

42. Koffler T et al. (2015), Genetics of hearing loss, *Otolaryngol Clin North Am* 48(6): 1041–61.

43. Novich SD, Eagleman DM (2015), Using space and time to encode vibrotactile information: Toward an estimate of the skin's achievable throughput, *Exp Brain Res* 233(10): 2777–88. Perrotta M, Asgeirsdottir T, Eagleman DM (2020). Deciphering sounds through patterns of vibration on the skin. (Under review). See also Neosensory.com. Could we have chosen something besides vibration? After all, skin contains several types of receptors that can be used to convey information, including vibration, temperature, itch, pain, and stretch. But we chose to concentrate on vibration because it's fast. Temperature has poor localization, and it's slow to perceive. Although stretch receptors may have promising spatial and temporal properties, you wouldn't want the long-term discomfort of skin stretching. And I probably don't need to say much about pain.

44. As a side note, think of the "accent" a deaf person has when she speaks. Is this some sort of speech impediment? No. Instead, a person who is totally deaf learns to vocalize by watching and copying the mouth movements of speaking people. While impersonating lip movements is a reasonably effective method, the problem is that the deaf observer cannot see what the speaker's tongue is doing. Try speaking a normal sentence while leaving your tongue resting on the bottom of your mouth. You'll sound exactly like a deaf speaker. One interesting opportunity is that people with our devices can overcome this hidden-tongue limitation. By comparing someone else's words with one's own vocalization, one can feel the difference—and thereby explore around the space of possibilities until the words sound the same.

45. See Alcorn S (1932), The Tadoma method, *Volta Rev* 34:195–98; Reed CM et al. (1985), Research on the Tadoma method of speech communication, *J Acoust Soc Am* 77:247–57.

46. Due to computational limitations in previous years, earlier attempts at sound-to-touch substitution relied on band-pass filtering audio and playing this output to the skin over vibrating solenoids. The solenoids operated at a fixed frequency of less than half the bandwidth of some of these band-passed channels, leading to aliasing noise. Further, multichannel versions of these devices have been limited in the number of possible vibration interfaces due to battery size and capacity constraints. Now computation is faster and cheaper. The desired mathematical transforms can be performed in real time, at essentially no expense, and without the need of custom integrated circuits. Current lithium-ion batteries support more vibration interfaces than could be used in prior tactile aids. For earlier attempts at sound-to-touch hearing devices, see Summers and Gratton (1995); Traunmuller (1980); Weisenberger et al. (1991); Reed and Delhorne (2003); Galvin et al. (2001). See also Cholewiak RW, Sherrick CE (1986), Tracking skill of a

deaf person with long-term tactile aid experience: A case study, *J Rehabil Res Dev* 23(2): 20–26.

47. Turchetti et al. (2011), Systematic review of the scientific literature on the economic evaluation of cochlear implants in paediatric patients, *Acta Otorhinolaryngol* 31(5): 311.

48. For people who already have a cochlear implant, wearing a vibrotactile device improves their ability to identify environmental sounds—such as a dog bark, a door knock, a car honk—by an average of 20 percent (data from internal Neosensory studies).

49. Danilov YP et al. (2007), Efficacy of electrotactile vestibular substitution in patients with peripheral and central vestibular loss, *J Vestib Res* 17(2–3): 119–30.

50. One more note on sensory substitution: Which approach is best for an individual—a retinal chip or sensory substitution—depends on the origin of the blindness. The retinal chip is ideal for people with diseases of photoreceptor degeneration (such as retinitis pigmentosa or age-related macular degeneration), because these pathologies leave the downstream visual system intact and capable of receiving signals from the implanted electrodes. Other forms of blindness can't use the retinal chip: if the problem is with other parts of the eye (say, a retinal detachment), or results from a problem later in the visual system (say, a tumor or tissue damage from a stroke), then the retinal chip will be of no use. In these cases, a sensory substitution or a direct plug-in to the brain, beyond the point of damage, would be the right tool for the job. Also, note that some investigators are exploring the combination of sensory-substitution devices with plug-ins (to retina or brain); the idea is that the sensory substitution helps the visual cortex to interpret the incoming information from a prosthesis. In other words, it serves as a guide for decoding the information.

51. For a firsthand description of the experience, watch Neil Harbisson's TED talk. Recent innovations include the extension of the eyeborg to encode saturation by volume. The eyeborg device has been translated into a chip that could, in theory, be implantable. For recent developments by other groups, see, for example, the Colorophone: Osinski D, Hjelme DR (2018), A sensory substitution device inspired by the human visual system, in 2018 *11th International Conference on Human System Interaction*. (HSI) (pp. 186-192). IEEE.

52. Off the success of the eyeborg and other projects, Harbisson and his partner have started the Cyborg Foundation, a nonprofit devoted to marrying technologies to human bodies.

53. Specifically, they spliced in a human photopigment. Jacobs GH et al. (2007), Emergence of novel color vision in mice engineered to express a human cone photopigment, *Science* 315(5819): 1723–25.

54. Mancuso K et al. (2009), Gene therapy for red-green colour blindness in adult primates, *Nature* 461:784–88. The research team injected a virus containing the red-detecting opsin gene behind the retina. After twenty weeks of practice, the monkeys could use the color vision to discriminate previously indistinguishable

colors. The monkey named Dalton was named after John Dalton, a British chemist who in 1794 became the first person to describe his own color blindness.

55. Jameson KA (2009), Tetrachromatic color vision, in *The Oxford Companion to Consciousness*, ed. P Wilken, T Bayne, and A Cleeremans (Oxford: Oxford University Press).

56. Crystalens accommodating lens (Bausch + Lomb). Cornell PJ (2011), Blue-violet subjective color changes after Crystalens implantation, *Cataract and Refractive Surgery Today*. For more on what it is like to experience a small extension into the UV range, see this blog post by Alek Komarnitsky, www.komar.org/faq/colorado -cataract-surgery-crystalens/ultra-violet-color-glow. By the way, note that most commercial "black lights" actually extend into the violet spectrum. So unless you've had an artificial lens implanted, that's probably why you detect a purplish light.

57. Ardouin J et al. (2012), FlyVIZ: A novel display device to provide humans with 360° vision by coupling catadioptric camera with HMD, in *Proceedings of the 18th ACM Symposium on Virtual Reality Software and Technology* (pp. 41-44); Guillermo AB et al. (2016), Enjoy 360° vision with the FlyVIZ, in *ACM SIGGRAPH 2016 Emerging Technologies* (New York: ACM), 6.

58. Wolbring G (2013), Hearing beyond the normal enabled by therapeutic devices: The role of the recipient and the hearing profession, *Neuroethics* 6:607.

59. Eagleman DM, Can we create new senses for humans?, TED talk, March 2015, ted .com. See also Hawkins, Blakeslee (2004).

60. Huffman, personal interview; Larratt, interview from Dvorsky G (2012), What does the future have in store for radical body modification?, *io9*. Note that Larratt had to have his magnets removed because the coating came off.

61. Nordmann GC, Hochstoeger T, Keays DA (2017), Magnetoreception—a sense without a receptor, *PLoS Biol* 15(10): e2003234.

62. Kaspar K et al. (2014), The experience of new sensorimotor contingencies by sensory augmentation, *Conscious Cogn* 28:47–63; Kärcher SM et al. (2012), Sensory augmentation for the blind, *Front Hum Neurosci* 6:37.

63. Nagel SK et al. (2005), Beyond sensory substitution—learning the sixth sense, *J Neural Eng* 2(4): R13.

64. Ibid.

65. See ibid. Also, note the interesting relationship of this to the blind people we met before who are better able to use their pinna (the external part of the ear) for localization than sighted people. In other words, sighted people have the same opportunity, but it's a tiny signal that lives below the surface of awareness. When one begins to need it, however, one can train up a weak signal and make good use of it.

Also, note that one won't have to wear a belt for long. At the end of 2018, scientists developed a thin electronic skin—essentially a little sticker on the hand—that indicates north. Cañón Bermúdez GS et al. (2018), Electronic-skin compasses for geomagnetic field-driven artificial magnetoreception and interactive electronics, *Nat Electron* 1:589–95.

66. Norimoto H, Ikegaya Y (2015), Visual cortical prosthesis with a geomagnetic compass restores spatial navigation in blind rats, *Curr Biol* 25(8): 1091–95.

67. It was perilous to fly without a visible horizon, a problem that was only solved with the invention of the gyroscope artificial horizon. In one of the World War II bombers, the copilot's seat was not precisely level, and that could cause them to go off course. There were training programs to counteract the "bottom effect."

68. By the way, Descartes came to the conclusion that he could never actually know whether this was all an illusion or not. But against that realization, he generated one of the most important steps in philosophy: he realized that *someone* is asking the question, so even if that someone is being manipulated by an evil demon, that someone still exists. *Cogito ergo sum:* I think, therefore I am. You may never be able to know if you're the victim of a demon, or a brain in a jar, but at least some *you* exists to vex about it. For the brain-in-a-vat argument, see Putnam H (1981), *Reason, Truth, and History* (New York: Cambridge University Press).

69. Neely RM et al. (2018), Recent advances in neural dust: Towards a neural interface platform, *Curr Opin Neurobiol* 50:64–71.

70. This line of questioning should not be confused with synesthesia, in which stimulation in one sense can trigger a sensation in another sense—such as sound triggering the experience of color. In synesthesia, one is well aware of the original stimulus, and also has an internal sense of something else. But in the main text I'm asking about actually *confusing* one sense for another. For more on synesthesia, see Cytowic RE, Eagleman DM (2009), *Wednesday Is Indigo Blue: Discovering the Brain of Synesthesia* (Cambridge, Mass.: MIT Press).

71. Eagleman DM (2018), We will leverage technology to create new senses, *Wired*.

72. O'Regan JK, Noë A (2001), A sensorimotor account of vision and visual consciousness, *Behav Brain Sci* 24(5): 939–73. Remember Bach-y-Rita's experiments with the blind people in the dental chair? The big improvements happened when the participants were able to establish the contingency between their actions and the resulting feedback: as they swung the camera around, the world changed in predictable ways. Whether biological or man-made, senses provide a way to actively explore the environment, matching a particular action to a specific change in the input. Bach-y-Rita (1972, 2004); Hurley, Noë (2003); Noë (2004).

73. Nagel et al. (2005).

5 How to Get a Better Body

1. Fuhr P et al. (1992), Physiological analysis of motor reorganization following lower limb amputation, *Electroencephalogr Clin Neurophysiol* 85(1): 53–60; Pascual-Leone A et al. (1996), Reorganization of human cortical motor output

maps following traumatic forearm amputation, *Neuroreport* 7:2068–70; Hallett M (1999), Plasticity in the human motor system, *Neuroscientist* 5:324–32; Karl A et al. (2001), Reorganization of motor and somatosensory cortex in upper extremity amputees with phantom limb pain, *J Neurosci* 21:3609–18.

2. Vargas CD et al. (2009), Re-emergence of hand-muscle representations in human motor cortex after hand allograft, *Proc Natl Acad Sci USA* 106(17): 7197–202.

3. Homeobox genes control the development of larger body structures. As an example, one of the first homeobox gene discoveries involved a mutation in fruit flies in which a pair of legs could grow where the antennae were supposed to be, and a reverse mutation that placed antennae where the legs should be. This happens because some genes act as a switch to turn on a cascade of other genes, and this is why many mutations involve the surprising appearance or disappearance of a full body "part"—for example, why some children are born with a tail. Mukhopadhyay B et al. (2012), Spectrum of human tails: A report of six cases, *J Indian Assoc Pediatr Surg* 17(1): 23–25.

4. Sommerville Q (2006), Three-armed boy "recovering well," *BBC News*, July 6, 2006.

5. Bongard J, Zykov V, Lipson H (2006), Resilient machines through continuous self-modeling, *Science* 314:1118–21; Pfeifer R, Lungarella M, Iida F (2007), Self-organization, embodiment, and biologically inspired robotics, *Science* 318(5853): 1088–93.

6. As a side note, a robot having a model of itself opens the door to its building other robots just like it, relying on trial and error to measure the distance between itself and what it's constructing.

7. Nicolelis M (2011), *Beyond Boundaries: The New Neuroscience of Connecting Brains with Machines—and How It Will Change Our Lives* (New York: St. Martin's Griffin).

8. Kennedy PR, Bakay RA (1998), Restoration of neural output from a paralyzed patient by a direct brain connection, *Neuroreport* 9:1707–11.

9. Hochberg LR et al. (2006), Neuronal ensemble control of prosthetic devices by a human with tetraplegia, *Nature* 442:164–71.

10. Jan's spinocerebellar degeneration is a rare disorder that ruins the communication between the brain and the muscles. For the science of Jan's case, see Collinger JL et al. (2013), High-performance neuroprosthetic control by an individual with tetraplegia, *Lancet* 381(9866): 557–64. For an overview of her treatment and its promises, see Eagleman DM (2016), *The Brain* (Edinburgh: Canongate Books), and Khatchadourian R (2018), Degrees of freedom, *New Yorker*.

11. Upton S (2014), What is it like to control a robotic arm with a brain implant?, *Scientific American*.

12. The most successful techniques require implanting electrodes directly into the cortex during neurosurgery, but less invasive techniques (for example, using electrodes on the outside of the head) are also in development.

13. Within each hemisphere, five areas will be implanted: the dorsal and ventral

aspects of the premotor cortices, the primary motor cortex, the primary somatosensory cortex, and the posterior parietal cortex. See www.WalkAgainProject.org for updates.

14. Bouton CE et al. (2016), Restoring cortical control of functional movement in a human with quadriplegia, *Nature* 533(7602): 247. The participant had a cervical spinal cord injury. The investigators used machine learning algorithms to learn how to best interpret the firestorm of neural activity and then send summary signals to a sophisticated electrical muscle stimulation system.

15. Iriki A, Tanaka M, Iwamura Y (1996), Attention-induced neuronal activity in the monkey somatosensory cortex revealed by pupillometrics, *Neurosci Res* 25(2): 173–81; Maravita A, Iriki A (2004), Tools for the body (schema), *Trends Cogn Sci* 8:79–86.

16. Velliste M et al. (2008), Cortical control of a prosthetic arm for self-feeding, *Nature* 453:1098–101.

   As a side note, we often think about robotic arms as being made of metal. But that won't be the case for long. "Soft robots" are built of stretchable rubbers and flexible plastics. Current research uses materials like cloth to build artificial fingers, octopus tentacles, worms, and so on. The shape changes by adjusting air pressure or by using electrical or chemical signals.

17. Fitzsimmons N et al. (2009), Extracting kinematic parameters for monkey bipedal walking from cortical neuronal ensemble activity, *Front Integr Neurosci* 3:3; Nicolelis M (2011), Limbs that move by thought control, *New Scientist* 210(2813): 26–27. See also Nicolelis's 2012 TEDMED talk: "A monkey that controls a robot with its thoughts. No, really."

18. See Nicolelis's book, *Beyond Boundaries*.

19. At least, Mother Nature never solved that problem *directly*. We could argue she solved the Bluetooth problem by evolving humans out of the primordial soup to do it for her.

20. Feeling that a body part is alien or bizarre is typically classified as asomatognosia, while denying a limb as one's own is a subset disorder known as somatoparaphrenia. See Feinberg T et al. (2010), The neuroanatomy of asomatognosia and somatoparaphrenia, *J Neurol Neurosurg Psychiatry* 81:276–81. See also Dieguez S, Annoni J-M (2013), Asomatognosia, in *The Behavioral and Cognitive Neurology of Stroke*, ed. O Goderfroy and J Bogousslavsky (Cambridge, U.K.: Cambridge University Press), 170. See also Feinberg TE (2001), *Altered Egos: How the Brain Creates the Self* (New York: Oxford University Press), and Arzy S et al. (2006), Neural mechanisms of embodiment: Asomatognosia due to premotor cortex damage, *Arch Neurol* 63:1022–25. Note that the jury is still out on whether all forms of asomatognosia are different flavors of the same disorder or instead fundamentally different disorders lumped under one label.

21. This last disorder, very rare, is known as misoplegia; see Pearce J (2007), Misoplegia, *Eur Neurol* 57:62–64.

22. Sacks OW (1984), *A Leg to Stand On* (New York: Harper & Row); Sacks OW

(1982), The leg, *London Review of Books*, June 17, 1982. See also Stone J, Perthen J, Carson AJ (2012), "A Leg to Stand On" by Oliver Sacks: A unique autobiographical account of functional paralysis, *J Neurol Neurosurg Psychiatry* 83(9): 864–67.

23. Simon M (2019), How I became a robot in London—from 5,000 miles away, *Wired*.

24. Herrera F et al. (2018), Building long-term empathy: A large-scale comparison of traditional and virtual reality perspective-taking, *PloS One* 13(10): e0204494; van Loon A et al. (2018), Virtual reality perspective-taking increases cognitive empathy for specific others, *PloS One* 13(8): e0202442. See also Bailenson J (2018), *Experience on Demand* (New York: W. W. Norton).

25. Won AS, Bailenson JN, Lanier J (2015), Homuncular flexibility: The human ability to inhabit nonhuman avatars, in *Emerging Trends in the Social and Behavioral Sciences,* ed. R Scott and M Buchmann (John Wiley & Sons), 1–6.

26. Won, Bailenson, Lanier (2015). See also Laha B et al. (2016), Evaluating control schemes for the third arm of an avatar, *Presence: Teleoperators and Virtual Environments* 25(2): 129–47.

27. Steptoe W, Steed A, Slater M (2013), Human tails: Ownership and control of extended humanoid avatars, *IEEE Trans Vis Comput Graph* 19:583–90.

28. Hershfield HE et al. (2011), Increasing saving behavior through age-progressed renderings of the future self, *JMR* 48(SPL): S23–37; Yee And et al. (2011), The expression of personality in virtual worlds, *Soc Psycho Pers Sci* 2(1): 5–12; Fox J et al (2009), Virtual experiences, physical behaviors: The effect of presence on imitation of an eating avatar, *Presence* 18(4): 294–303.

29. DeCandido K (1997), "Arms and the man," in *Untold Tales of Spider-Man,* ed. S Lee and K Busiek (New York: Boulevard Books).

30. Wetzel F (2012), Dad who lost arm gets new lease of life with most hi-tech bionic hand ever, *Sun*.

31. Eagleman DM (2011), in "20 predictions for the next 25 years," *Observer*, Jan. 2, 2011.

6 Why Mattering Matters

1. There might have been genetic advantages as well; it's very difficult to know. But there exist no genes that code for chess directly, so the years of training were certainly necessary.

2. Schweighofer N, Arbib MA (1998), A model of cerebellar metaplasticity, *Learn Mem* 4(5): 421–28.

3. As a strange side note, I originally heard this story about Perlman, but I have now found it variously attributed online to Perlman, Fritz Kreisler, Isaac Stern, and several other musicians. Whoever said it, everyone wants credit for the good line.

4. Elbert T et al. (1995), Increased finger representation of the fingers of the left hand in string players, *Science* 270:305–6; Bangert M, Schlaug G (2006), Specialization

of the specialized in features of external human brain morphology, *Eur J Neurosci* 24:1832–34. If you look at the gyrus (the elevated ridge on the landscape of the brain) in nonmusicians, you'll see it runs generally straight; the same gyrus in musicians has a strange, puckering detour. Note that while a violinist's left hand is doing all the detailed work, he shows the omega sign in the right hemisphere; this is because the left hand is represented on the right side of the brain.

5. This was first demonstrated in monkeys trained on one of two different tasks: retrieving a small object from a well, or turning a large key. The first task required skilled, fine use of the fingers, while the second relied on wrist and forearm use. After the monkeys trained on the first task, the cortical representation for the fingers progressively usurped more territory while the wrist and forearm representation shrank. In contrast, if the monkeys trained on the key-turning task, the amount of neural territory devoted to the wrist and forearm expanded. Nudo RJ et al. (1996), Use-dependent alterations of movement representations in primary motor cortex of adult squirrel monkeys, *J Neurosci* 16(2): 785–807.

6. Karni A et al. (1995), Functional MRI evidence for adult motor cortex plasticity during motor skill learning, *Nature* 377:155–58.

7. Draganski B et al. (2004), Neuroplasticity: Changes in grey matter induced by training, *Nature* 427(6972): 311–12; Driemeyer J et al. (2008), Changes in gray matter induced by learning—revisited, *PLoS One* 3(7): e2669; Boyke J et al. (2008), Training-induced brain structure changes in the elderly, *J Neurosci* 28(28): 7031–35; Scholz J et al. (2009), Training induces changes in white-matter architecture, *Nat Neurosci* 12(11): 1370–71. The hypothesis is that density increases in the gray matter seen within a week of training are probably due to an increased size of synapses or cell bodies, while increased volume on the longer term (months) may reflect the birth of new neurons, especially in the hippocampus.

8. Eagleman DM (2011), *Incognito: The Secret Lives of the Brain* (New York: Pantheon).

9. Iriki A, Tanaka M, Iwamura Y (1996), Attention-induced neuronal activity in the monkey somatosensory cortex revealed by pupillometrics, *Neurosci Res* 25(2): 173–81; Maravita A, Iriki A (2004), Tools for the body (schema), *Trends Cogn Sci* 8:79–86.

10. Draganski B et al. (2006), Temporal and spatial dynamics of brain structure changes during extensive learning, *J Neurosci* 26(23): 6314–17.

11. Ilg R et al. (2008), Gray matter increase induced by practice correlates with task-specific activation: A combined functional and morphometric magnetic resonance imaging study, *J Neurosci* 28(16): 4210–15.

12. Maguire EA et al (2000), Navigation-related structural change in the hippocampi of taxi drivers, *Proc Natl Acad Sci USA* 97(8): 4398–403. See also Maguire EA, Frackowiak RS, Frith CD (1997), Recalling routes around London: Activation of the right hippocampus in taxi drivers, *J Neurosci* 17(18): 7103–10.

13. Kuhl PK (2004), Early language acquisition: Cracking the speech code, *Nat Rev Neurosci* 5:831–43.

14. These studies were originally performed with monkeys. In one study, a monkey was exposed to simultaneous auditory and tactile stimulation. If the demands of the task required him to attend to the touch, his somatosensory cortex showed plastic changes while his auditory cortex did not. If he instead was directed to attend to the auditory stimulus, the opposite happened. See Recanzone GH et al. (1993), Plasticity in the frequency representation of primary auditory cortex following discrimination training in adult owl monkeys, *J Neurosci* 13(1): 87–103; Jenkins WM et al. (1990), Functional reorganization of primary somatosensory cortex in adult owl monkeys after behaviorally controlled tactile stimulation, *J Neurophysiol* 63(1): 82–104; Bavelier D, Neville HJ (2002), Cross-modal plasticity: Where and how?, *Nat Rev Neurosci* 3(6): 443.

15. Taub E, Uswatte G, Pidikiti R (1999), Constraint-induced movement therapy: A new family of techniques with broad application to physical rehabilitation, *J Rehabil Res Dev* 36(3): 1–21; Page SJ, Boe S, Levine P (2013), What are the "ingredients" of modified constraint induced therapy? An evidence-based review, recipe, and recommendations, *Restor Neurol Neurosci* 31:299–309.

16. Teng S, Whitney D (2011), The acuity of echolocation: Spatial resolution in the sighted compared to expert performance, *J Vis Impair Blind* 105(1): 20.

17. Further, nations have a remarkably flexible body plan. As new territories are acquired, they become part of a nation's limbs and consciousness. Diplomats are put into position, and military bases are set up. The location becomes a tele-limb of the nation, and just as with any limb the body can control, the outposts become part of the selfhood of the country. Also, note how readily governments wrap around new inputs: as technology changes, agencies and legislation adjust to match its shape.

18. Neuroscientists use the term "neurotransmitter" to refer to a chemical messenger released from a neuron at a specialized junction, where it communicates with another cell with high specificity. A neuromodulator, in contrast, is a chemical messenger that affects a larger population of neurons (or other cell types) and tends to have broader effects. Note that a particular chemical can act as a transmitter or a modulator in different circumstances: acetylcholine acts as a transmitter in the periphery (when communicating with muscle cells), but as a modulator in the central nervous system.

19. Bakin JS, Weinberger NM (1996), Induction of a physiological memory in the cerebral cortex by stimulation of the nucleus basalis, *Proc Natl Acad Sci USA* 93:11219–24.

20. Neurons that release acetylcholine are called cholinergic, and these neurons exist almost entirely in the basal forebrain, a subcortical collection of structures that project to the cortex. It has many effects in the central nervous system, including altering the excitability of neurons, modulating the presynaptic release of neurotransmitters, and coordinating the firing of small populations of neurons. See Picciotto MR, Higley MJ, Mineur YS (2012), Acetylcholine as a neuromodula-

tor: Cholinergic signaling shapes nervous system function and behavior, *Neuron* 76(1): 116–29; Gu Q (2003), Contribution of acetylcholine to visual cortex plasticity, *Neurobiol Learn Mem* 80:291–301; Richardson RT, DeLong MR (1991), Electrophysiological studies of the functions of the nucleus basalis in primates, *Adv Exp Med Biol* 295:233–52; Orsetti M, Casamenti F, Pepeu G (1996), Enhanced acetylcholine release in the hippocampus and cortex during acquisition of an operant behavior, *Brain Res* 724:89–96. Note that many neuromodulators transiently change the balance of excitation and inhibition; this had led to one hypothesis that disinhibition is one mechanism by which neuromodulation enables long-term synaptic modifications.

21. Hasselmo ME (1995), Neuromodulation and cortical function: Modeling the physiological basis of behavior, *Behav Brain Res* 67:1–27.

22. This was first demonstrated in adult rats some decades ago. If the rats are exposed to particular auditory tones, the tones alone do not result in any significant changes in cortical representation. But if a particular tone was paired with stimulation of the cholinergic nucleus basalis, the cortical representation for that tone expanded. Kilgard MP, Merzenich MM (1998), Cortical map reorganization enabled by nucleus basalis activity, *Science* 279:1714–18. For a review of the science in both rats and people, see Weinberger NM (2015), New perspectives on the auditory cortex: Learning and memory, *Handb Clin Neurol* 129:117–47.

23. Bear MF, Singer W (1986), Modulation of visual cortical plasticity by acetylcholine and noradrenaline, *Nature* 320:172–76; Sachdev RNS et al. (1998), Role of the basal forebrain cholinergic projection in somatosensory cortical plasticity, *J Neurophysiol* 79:3216–28.

24. Conner JM et al. (2003), Lesions of the basal forebrain cholinergic system impair task acquisition and abolish cortical plasticity associated with motor skill learning, *Neuron* 38:819–29.

25. For the full interview, in which, by the way, Asimov foresaw the internet years ahead of its blossoming, search for the video on YouTube.

26. Brandt A, Eagleman DM (2017), *The Runaway Species* (New York: Catapult).

7 Why Love Knows Not Its Own Depth Until the Hour of Separation

1. Eagleman DM (2001), Visual illusions and neurobiology, *Nat Rev Neurosci* 2(12): 920–26.

2. Pelah A, Barlow HB (1996), Visual illusion from running, *Nature* 381(6580): 283; Zadra JR, Proffitt DR (2016), Optic flow is calibrated to walking effort, *Psychon Bull Rev* 23(5): 1491–96.

3. This illusion is known as the McCullough effect, named after Celeste McCullough, who discovered it in 1965. McCullough C (1965), Color adaptation of edge-detectors in the human visual system, *Science* 149:1115–16. Note that this illusion

won't be possible for people who are color-blind. This contingent aftereffect works not only with oriented lines and color but also between movement and color, spatial frequency and color, and more.

4. Jones PD, Holding DH (1975), Extremely long-term persistence of the McCollough effect, *J Exp Psychol Hum Percept Perform* 1(4): 323–27.

5. The big movements of the eyes are called saccades, and the smaller jitters in between are called micro-saccades.

6. This is called entoptic vision, referring to effects coming from the eye itself (*entoptical*), as opposed to visual illusions, which arise because of the brain's interpretation. For background on illusions arising from within the eye, see Tyler CW (1978), Some new entoptic phenomena, *Vision Res* 18(12): 1633–39.

7. This was first noted by Jan Purkinje in 1823 and hence the image of the blood vessels in one's own eye is known as Purkinje's Tree. See Purkyně J (1823), *Beiträge zur Kenntniss des Sehens in subjectiver Hinsicht*, in *Beobachtungen und Versuche zur Physiologie der Sinne* (Prague: In Commission der J. G. Calve'schen Buchhandlung).

8. Stetson C et al. (2006), Motor-sensory recalibration leads to an illusory reversal of action and sensation, *Neuron* 51(5): 651–59.

9. I think this is an interesting idea to keep in science's back pocket: Are there things that seem to appear, that actually represent something disappearing?

10. Kamin LJ (1969), Predictability, surprise, attention, and conditioning, in *Punishment and Aversive Behavior*, ed. BA Campbell and RM Church (New York: Appleton-Century-Crofts), 279–96; Bouton ME (2007), *Learning and Behavior: A Contemporary Synthesis* (Sunderland, Mass.: Sinauer).

11. This method of gradient climbing is known as klinokinesis.

12. Rods and cones cover only four orders of magnitude of illuminance in the dark, but in continuous ambient light can cover much more. Due to a complex variety of mechanisms, photoreceptors avoid saturation and respond to an increased flux of photons by adjusting the amplification factors (and rates of recovery) of their molecular cascades. Some examples: changing the lifetime of molecules in their biochemically active state, changing the availability of binding proteins nearby, using other molecules to increase the lifetime of the activated complexes, and changing the affinity of channels for the ligands that bind to them. On a larger scale, photoreceptors can join forces with one another thanks to horizontal cells, which modulate the connections (called gap junctions) to change the way the photoreceptors interact. See Arshavsky VY, Burns ME (2012), Photoreceptor signaling: Supporting vision across a wide range of light intensities, *J Biol Chem* 287(3): 1620–26; Chen J et al. (2010), Channel modulation and the mechanism of light adaptation in mouse rods, *J Neurosci* 30(48): 16232–40; Diamond JS (2017), Inhibitory interneurons in the retina: Types, circuitry, and function, *Annu Rev Vis Sci* 3:1–24; O'Brien J, Bloomfield SA (2018), Plasticity of retinal gap junctions: Roles in synaptic physiology and disease, *Annu Rev Vis Sci* 4:79–100; Demb JB, Singer JH (2015), Functional circuitry of the retina, *Annu Rev Vis Sci* 1:263–89.

8 Balancing on the Edge of Change

1. Muckli L, Naumer MJ, Singer W (2009), Bilateral visual field maps in a patient with only one hemisphere, *Proc Natl Acad Sci* 106(31): 13034–39. Note that Alice also had an unusually small right eye that was mostly blind. See the supplemental material from this paper for more background on her.

2. Udin SH (1977), Rearrangements of the retinotectal projection in *Rana pipiens* after unilateral caudal half-tectum ablation, *J Comp Neurol* 173:561–82.

3. Constantine-Paton M, Law MI (1978), Eye-specific termination bands in tecta of three-eyed frogs, *Science* 202:639–41; Law MI, Constantine-Paton M (1981), Anatomy and physiology of experimentally produced striped tecta, *J Neurosci* 1:741–59.

4. Why do the territories alternate in stripes instead of an even blending? Computer models show this falls naturally out of Hebbian competition between the incoming axons from the different eyes. A central model of the formation of the stripes seen in ocular dominance columns was proposed in the 1980s (Miller KD, Keller JB, Stryker MP [1989], Ocular dominance column development: Analysis and simulation, *Science* 245[4918]: 605–15). This model has since been built upon with the addition of many other physiologically realistic features.

5. Attardi DG, Sperry RW (1963), Preferential selection of central pathways by regenerating optic fibers, *Exp Neurol* 7(1): 46–64.

6. Basso A et al. (1989), The role of the right hemisphere in recovery from aphasia: Two case studies, *Cortex* 25:555–66. In recent decades, researchers have been able to witness the transfer in action with neuroimaging. See Heiss WD, Thiel A (2006), A proposed regional hierarchy in recovery of post-stroke aphasia, *Brain Lang* 98:118–23; Pani E et al. (2016), Right hemisphere structures predict post-stroke speech fluency, *Neurology* 86:1574–81; Xing S et al. (2016), Right hemisphere grey matter structure and language outcome in chronic left-hemisphere stroke, *Brain* 139:227–41. The amount of "right shifting" that is clinically observed differs from patient to patient, for reasons that are still being investigated.

7. Wiesel TN, Hubel DH (1963), Single-cell responses in striate cortex of kittens deprived of vision in one eye, *J Neurophysiol* 26:1003–17; Gu Q (2003), Contribution of acetylcholine to visual cortex plasticity, *Neurobiol Learn Mem* 80:291–301; Hubel DH, Wiesel TN (1965), Binocular interaction in striate cortex of kittens reared with artificial squint, *J Neurophysiol* 28:1041–59. Such experiments were originally performed with kittens and monkeys; later technologies confirmed (not surprisingly) that exactly the same lessons apply to understanding the human visual cortex.

8. Note this is somewhat analogous to the strategy of constraint therapy in stroke patients, in which the good arm is put in a sling.

9. This is called a relational map: the hand will be represented near the elbow, which will be represented near the shoulder, irrespective of how much or how little territory is available.

10. In the years since Levi-Montalcini's initial discovery, a whole family of other neu-rotrophic factors have been discovered; they all have in common the property that they stimulate the survival and development of neurons. More generally, neuro-trophins belong to a class of secreted proteins known as growth factors. See Spedding M, Gressens P (2008), Neurotrophins and cytokines in neuronal plasticity, *Novartis Found Symp* 289:222–33.

11. Zoubine MN et al. (1996), A molecular mechanism for synapse elimination: Novel inhibition of locally generated thrombin delays synapse loss in neonatal mouse muscle, *Dev Biol* 179:447–57.

12. Sanes JR, Lichtman JW (1999), Development of the vertebrate neuromuscular junction, *Annu Rev Neurosci* 22:389–442.

13. Consider what happened in 1933 in Germany, when almost all of the elected representatives to the Reichstag were from either far-left parties (such as the Communists) or far-right parties (such as the Nazis). Although the balance was represented by the extremes, it was balance nonetheless. But in August 1934, following President Paul von Hindenburg's death, Adolf Hitler declared himself *Führer und Reichskanzler* and passed laws by decree. Not surprisingly, the first laws rounded up his Communist opponents into concentration camps, and the massive imbalance triggered disaster for millions.

14. Yamahachi H et al. (2009), Rapid axonal sprouting and pruning accompany func-tional reorganization in primary visual cortex, *Neuron* 64(5): 719–29; Buonomano DV, Merzenich MM (1998), Cortical plasticity: From synapses to maps, *Annu Rev Neurosci* 21:149–86; Pascual-Leone A, Hamilton R (2001), The metamodal organization of the brain, in *Vision: From Neurons to Cognition*, ed. C Casanova and M Ptito (New York: Elsevier Science), 427–45; Pascual-Leone A et al. (2005), The plastic human brain cortex, *Annu Rev Neurosci* 28:377–401; Merzenich MM et al. (1984), Somatosensory cortical map changes following digit amputation in adult monkeys, *J Comp Neurol* 224:591–605; Pons TP et al. (1991), Massive cortical reorganization after sensory deafferentation in adult macaques, *Science* 252:1857–60; Sanes JN, Donoghue JP (2000), Plasticity and primary motor cortex, *Annu Rev Neurosci* 23(1): 393–415.

15. Jacobs KM, Donoghue JP (1991), Reshaping the cortical motor map by unmasking latent intracortical connections, *Science* 251(4996): 944–7; Tremere L et al. (2001), Expansion of receptive fields in raccoon somatosensory cortex in vivo by GABA-A receptor antagonism: Implications for cortical reorganization, *Exp Brain Res* 136(4): 447–55.

16. This mechanism sharpens the boundaries between regions. For example, see Tre-mere et al. (2001).

17. Weiss T et al. (2004), Rapid functional plasticity in the primary somatomotor cortex and perceptual changes after nerve block, *Eur J Neurosci* 20:3413–23.

18. Bavelier D, Neville HJ (2002), Cross-modal plasticity: Where and how?, *Nat Rev Neurosci* 3:443–52.

19. Eckert MA et al. (2008), A cross-modal system linking primary auditory and

visual cortices: Evidence from intrinsic fMRI connectivity analysis, *Hum Brain Mapp* 29(7):848–57; Petro LS, Paton AT, Muckli L (2017), Contextual modulation of primary visual cortex by auditory signals, *Philos Trans R Soc B Biol Sci* 372(1714): 20160104.

20. Pascual-Leone et al. (2005).

21. Darian-Smith C, Gilbert CD (1994), Axonal sprouting accompanies functional reorganization in adult cat striate cortex, *Nature* 368:737–40; Florence SL, Taub HB, Kaas JH (1998), Large-scale sprouting of cortical connections after peripheral injury in adult macaque monkeys, *Science* 282:1117–21. A great deal of attention has been paid to changes within cortex, although again long-term changes in the thalamus may also contribute to slow, larger changes in cortical structure. See Jones EG (2000), Cortical and subcortical contributions to activity-dependent plasticity in primate somatosensory cortex, *Annu Rev Neurosci* 23:1–37; Buonomano, Merzenich (1998). For students of the next generation: an open biological question remains *how* to couple the fast changes (unmasking) to longer-term changes (growth of new axons).

22. Merlo LM et al. (2006), Cancer as an evolutionary and ecological process, *Nat Rev Cancer* 6(12): 924–35; Sprouffske K et al. (2012), Cancer in light of experimental evolution, *Curr Biol* 22(17): R762–R771; Aktipis CA et al. (2015), Cancer across the tree of life: Cooperation and cheating in multicellularity, *Philos Trans R Soc B Biol Sci* 370(1673).

9 Why Is It Harder to Teach Old Dogs New Tricks?

1. Teuber HL (1975), Recovery of function after brain injury in man, in *Outcome of Severe Damage to the Central Nervous System*, ed. R Porter and DW Fitzsimmons (Amsterdam: Elsevier), 159–90.

2. Young brains have high levels of cholinergic transmitters, but not the other, inhibitory transmitters, which become available only later; this gives them generalized plasticity. Adult brains, in contrast, actively inhibit change wherever it shouldn't happen. That is, the cholinergic effects in an adult brain are modified by inhibitory transmitters—which make most areas less plastic or not plastic, so that the brain only changes where necessary. See Gopnik A, Schulz L (2004), Mechanisms of theory formation in young children, *Trends Cogn Sci* 8:371–77; Schulz LE, Gopnik A (2004), Causal learning across domains, *Dev Psychol* 40:162–76. Because young brains allow for global change, the scientist Alison Gopnik calls babies the "research and development" department of the human species.

3. Gopnik A (2009), *The Philosophical Baby: What Children's Minds Tell Us About Truth, Love, and the Meaning of Life* (New York: Farrar, Straus & Giroux).

4. This description is adapted from Coch D, Fischer KW, Dawson G (2007), Dynamic development of the hemispheric biases in three cases: Cognitive/hemispheric cycles, music, and hemispherectomy, in *Human Behavior, Learning, and*

the Developing Brain (New York: Guilford), 94–97. Remarkably, this surgery has been performed successfully in adults, but it's uncommon and typically results in worse outcomes. See Schramm J et al. (2012), Seizure outcome, functional outcome, and quality of life after hemispherectomy in adults, *Acta Neurochir* 154(9): 1603–12.

5. The sensitive period is sometimes called the critical period.

6. Petitto LA, Marentette PF (1991), Babbling in the manual mode: Evidence for the ontogeny of language, *Science* 251:1493–96.

7. Lenneberg E (1967), *Biological Foundations of Language* (New York: Wiley); Johnson JS, Newport EL (1989), Critical period effects in second language learning: The influence of maturational state on the acquisition of English as a second language, *Cogn Psychol* 21:60–99. Note there is some controversy about whether plasticity is the explanation for everything in second-language acquisition; after all, sometimes adults can learn second languages faster than infants due to greater cognitive maturity, learning experience, and other psychological and social factors (see Newport [1990] and Snow, Hoefnagel-Hoehle [1978]). However, irrespective of the skill at learning a second language, native-like pronunciation in a foreign language (that is, accent) remains harder to achieve well for older learners. Asher J, Garcia R (1969), The optimal age to learn a foreign language, *Mod Lang J* 53(5): 334–341.

8. See Berman N, Murphy EH (1981), The critical period for alteration in cortical binocularity resulting from divergent and convergent strabismus, *Dev Brain Res* 2(2): 181–202. Having misaligned eyes is known colloquially as wall-eyed or cross-eyed, and technically as strabismus.

9. Amedi A et al. (2003), Early "visual" cortex activation correlates with superior verbal-memory performance in the blind, *Nat Neurosci* 6:758–66.

10. Voss P et al. (2006), A positron emission tomography study during auditory localization by late-onset blind individuals, *Neuroreport* 17(4): 383–88; Voss P et al. (2008), Differential occipital responses in early- and late-blind individuals during a sound-source discrimination task, *Neuroimage* 40(2): 746–58.

11. Merabet LB et al. (2005), What blindness can tell us about seeing again: Merging neuroplasticity and neuroprostheses, *Nat Rev Neurosci* 6(1): 71.

12. In other words, the studies found that while the auditory cortex came to look like a visual cortex, the new connections retained some features typical of an auditory cortex. For example, the new visual fields showed higher precision along the left-right axis than along the up-down axis, and it is thought to be because the auditory cortex normally maps frequencies in order along the left-right axis.

13. Persico N, Postlewaite A, Silverman D (2004), The effect of adolescent experience on labor market outcomes: The case of height, *J Polit Econ* 112(5): 1019–53. See also Judge TA, Cable DM (2004), The effect of physical height on workplace success and income: Preliminary test of a theoretical model, *J Appl Psychol* 89(3): 428–41.

14. Smirnakis et al. (2005), Lack of long-term cortical reorganization after macaque retinal lesions, *Nature* 435(7040): 300. This study was done with adult macaque

monkeys (older than four years old); the same lessons presumably hold for humans.

15. As one example of hundreds, recall that if you start to use a rake to grab your food, your sensory and motor cortices will quickly readjust to incorporate the rake into your body plan, even when you're an adult. See Iriki A, Tanaka M, Iwamura Y (1996), Attention-induced neuronal activity in the monkey somatosensory cortex revealed by pupillometrics, *Neurosci Res* 25(2): 173–81; Maravita A, Iriki A (2004), Tools for the body (schema), *Trends Cogn Sci* 8:79–86.

16. Chalupa LM, Dreher B (1991), High precision systems require high precision "blueprints": A new view regarding the formation of connections in the mammalian visual system, *J Cogn Neurosci* 3(3): 209–19; Neville H, Bavelier D (2002), Human brain plasticity: Evidence from sensory deprivation and altered language experience, *Prog Brain Res* 138:177–88.

17. Haldane JBS (1932), *The Causes of Evolution* (New York: Longmans, Green). Via S, Lande R (1985), Genotype-environment interaction and the evolution of phenotypic plasticity, *Evolution* 39:505–22; Via S, Lande R (1987), Evolution of genetic variability in a spatially heterogeneous environment: Effects of genotype-environment interaction, *Genet Res* 49:147–56.

18. Snowdon DA (2003), Healthy aging and dementia: Findings from the Nun Study, *Ann Intern Med* 139(5, pt. 2): 450–54.

19. The most critical thing as one ages is to figure out how to avoid entrenchment. As an analogy, the worst thing that can happen to a scientist is falling into the same way of looking at a problem or at a field. This may explain the surprising advantage of polymaths: people like Benjamin Franklin, who excel across many different fields. Because they constantly put themselves in new territory, they can avoid the trap of becoming fixed in a single way of thinking.

10 Remember When

1. Ribot T (1882), *Diseases of the Memory: An Essay in the Positive Psychology* (New York: D. Appleton).

2. Hawkins RD, Clark GA, Kandel ER (2006), Operant conditioning of gill withdrawal in aplysia, *J Neurosci* 26:2443–48.

3. Hebb DO (1949), *The Organization of Behavior: A Neuropsychological Theory* (New York: Wiley). As Hebb put it, "When an axon of cell *A* is near enough to excite a cell *B* and repeatedly or persistently takes part in firing it, some growth process or metabolic change takes place in one or both cells such that *A*'s efficiency, as one of the cells firing *B*, is increased." Although neuroscientists often refer to a synapse between A and B, keep in mind that A is also connecting to C through Z and to about ten thousand other neurons. The key is that each one of these synapses can change its strength individually, strengthening some conversations and weakening others.

4. Bliss TV, Lømo T (1973), Long-lasting potentiation of synaptic transmission in the dentate area of the anaesthetized rabbit following stimulation of the perforant path, *J Physiol (London)* 232(2): 331–56. At the level of the submicroscopic, tiny channels in the membrane sensitive to a particular chemical signal (known as NMDA receptors) act as *coincidence detectors,* responding when two connected neurons fire within a small time window. Many postsynaptic membranes contain NMDA glutamate receptors as well as non-NMDA glutamate receptors. During normal low-frequency stimulation, only non-NMDA channels will open, because naturally occurring magnesium ions block the NMDA channels. But high-frequency presynaptic input resulting in depolarization of the postsynaptic membrane displaces the magnesium ions, making the NMDA receptor sensitive to subsequent release of glutamate. In this way, the NMDA-R can act as a coincidence detector, sensing coincidence of pre- and postsynaptic activity. Thus, NMDA synapses appear to be the quintessential Hebbian synapses and have been eyed as the keys for the storage of associations. Moreover, the fact that NMDA-Rs have a particularly high permeability for calcium allows them to induce a second-messenger system that can eventually talk to the genome and result in long-term structural changes to the postsynaptic cell. In most types of neurons, the NMDA-R is crucial for induction of long-term potentiation (LTP). An animal can be taught a behavioral task, but with the infusion of chemicals that act like NMDA, the ability to remember the specifics of the task seems to disappear. However, note that the NMDA-R is only necessary for *induction,* while other mechanisms underlie the *maintenance* of the changes; most generally, new protein synthesis is required at the nucleus of the cell. An animal can be trained to associate two stimuli (say, pairing a shock with a light), but if protein synthesis is blocked, the animal can form short-term but not long-term memory. In most cases, LTP is induced only when the activity in the postsynaptic cell (depolarization) is associated with activity in the presynaptic cell. Post- or presynaptic activity alone is ineffective. Additionally, LTP is specific to the particular synapse being stimulated, which means that each individual synapse on a cell could, in principle, strengthen or weaken according to its own personal history.

5. Regarding the role of the synapse in memory, see Nabavi S et al. (2014), Engineering a memory with LTD and LTP, *Nature* 511:348–52; Bailey CH, Kandel RR (1993), Structural changes accompanying memory storage, *Annu Rev Physiol* 55:397–426.

6. Hopfield J (1982), Neural networks and physical systems with emergent collective computational abilities, *Proc Natl Acad Sci USA* 9:2554. Because each unit has many connections (synapses) to neighbors, a unit can be involved in many different associations at different times.

7. While Hebb's rule is useful for forming associations, one of its theoretical shortcomings is that it is insensitive to the *order* of events. Experiments have long shown that animals are strictly sensitive to the order of sensory inputs; for example, Pavlov's dog will not learn an association if the meat is presented before the bell. Likewise, animals develop strong aversion to a tasty food following a single

experience of nausea after eating it, but reversing the order (nausea and then the food) does not lead to an aversion. There may be a parallel to this at the biophysical level: changes in synaptic strength depend on the order of pre- and postsynaptic activity. If an input from A precedes the firing of neuron B, then the synapse is strengthened. If an input from A comes after cell B has fired, the synapse is weakened. This learning rule is commonly called spike-timing-dependent plasticity, or a temporally asymmetric Hebbian rule, and it suggests that spike timing matters. Specifically, the temporally asymmetric rule strengthens connections that are predictive: if A consistently fires before B, it can be viewed as a successful prediction and will be strengthened. See Rao RP, Sejnowski TJ (2003), Self-organizing neural systems based on predictive learning, *Philos Transact A Math Phys Eng Sci* 361(1807): 1149–75.

8. The core concepts that underlie deep learning are well over thirty years old. See Rumelhart DE, Hinton GE, Williams RJ (1988), Learning representations by back-propagating errors, *Cognitive Modeling* 5(3): 1. See also the work of Yann LeCun, Yoshua Bengio, and Jürgen Schmidhuber for related key developments around that same time.

9. Carpenter GA, Grossberg S (1987), Discovering order in chaos: Stable self-organization of neural recognition codes, *Ann NY Acad Sci* 504:33–51.

10. Bakin JS, Weinberger NM (1996), Induction of a physiological memory in the cerebral cortex by stimulation of the nucleus basalis, *Proc Natl Acad Sci USA* 93:11219–24; Kilgard MP, Merzenich MM (1998), Cortical map reorganization enabled by nucleus basalis activity, *Science* 279:1714–18.

11. Note that Molaison's deficits were totally unexpected, because removal of the medial temporal lobe (the hippocampus and surrounding regions) on *one* side had been known to be a safe procedure for some time. For a summary of his life and clinical case, see Corkin S (2013), *Permanent Present Tense: The Unforgettable Life of the Amnesic Patient, HM* (New York: Basic Books).

12. Zola-Morgan SM, Squire LR (1990), The primate hippocampal formation: Evidence for a time-limited role in memory storage, *Science* 250(4978): 288–90.

13. Eichenbaum H (2004), Hippocampus: Cognitive processes and neural representations that underlie declarative memory, *Neuron* 44(1): 109–20. See also Frankland PW et al. (2004), The involvement of the anterior cingulate cortex in remote contextual fear memory, *Science* 304(5672): 881–83.

14. Pasupathy A, Miller EK (2005), Different time courses of learning-related activity in the prefrontal cortex and striatum, *Nature* 433(7028): 873–76. See also Ravel S, Richmond BJ (2005), Where did the time go?, *Nat Neurosci* 8(6): 705–7.

15. Lisman J et al. (2018), Memory formation depends on both synapse-specific modifications of synaptic strength and cell-specific increases in excitability, *Nat Neurosci* 12:1; Martin SJ, Grimwood PD, Morris RG (2000), Synaptic plasticity and memory: An evaluation of the hypothesis, *Annu Rev Neurosci* 23:649–711; Shors TJ, Matzel LD (1997), Long-term potentiation: What's learning got to do with it?, *Behav Brain Sci* 20(4): 597–655. As regards LTP and LTD, much is still unknown

about how the intracellular context of neurons determines how synapses will change: not all synapses behave alike. It was initially hoped that the details of the stimulation protocols would determine the outcome: a high firing rate will strengthen a synapse, and a low firing rate will weaken it. But as the experimental studies took off, some investigators who found that a cell did not depress when given the "right" stimulus tended to throw away those data, under the assumption that the cell was "sick." A more sober look at the data reveals that the synaptic rules for change pivot on other factors inside the cell, most of which remain to be identified. See Perrett SP et al. (2001), LTD induction in adult visual cortex: Role of stimulus timing and inhibition, *J Neurosci* 21(7): 2308–19.

16. Draganski B et al. (2004), Neuroplasticity: Changes in grey matter induced by training, *Nature* 427(6972): 311–12.

17. For example, newly branching axons or dendrites, or the birth of new glial cells or neurons.

18. Boldrini M et al. (2018), Human hippocampal neurogenesis persists throughout aging, *Cell Stem Cell* 22(4): 589–99; Gould et al. (1999), Neurogenesis in the neocortex of adult primates, *Science* 286(5439): 548–52; Eriksson et al. (1998), Neurogenesis in the adult human hippocampus, *Nat Med* 4(11): 1313.

    Dogma since the 1960s advised that mammals are born with a fixed number of neurons: the number can diminish with age, but can never increase. But with increased resolution in techniques, we now know that the hippocampus is churning out thousands of new neurons each day, in animals from mice to humans. It's only through a historical mistake that this finding of neurogenesis comes as a surprise; after all, new cell growth characterizes every other part of the body, and we've known for a long time that bird brains do this—in fact, every time they need to learn a new song: Nottebohm F (2002), Neuronal replacement in adult brain, *Brain Res Bull* 57(6): 737–49. As a point of historical interest, neurogenesis in the mammalian brain was suspected for a long time, but ignored; see Altman J (1962), Are new neurons formed in the brains of adult mammals?, *Science* 135(3509): 1127–28.

19. Gould E et al. (1999), Learning enhances adult neurogenesis in the adult hippocampal formation, *Nat Neurosci* 2:260–65. So why wouldn't existing memories be disrupted by these interlopers? If new cells can insinuate themselves into the fabric of the cortex without corrupting stored old memories, something about the connectome paradigm needs revamping. One speculation is that synapses, perhaps by virtue of the turnover of their constituent molecules, are not reliable repositories for learned, long-term information (Nottebohm [2002]; Bailey, Kandel [1993]). Instead, the final biophysical change requires a whole new neuron. In this speculative framework, the storage of a memory involves the activation of a set of genes that leads to cell differentiation. The irreversibility of the cell division is just what one would want for long-term memory storage on a longer timescale. I want to flag this idea as speculative, mostly because so much remains to be understood about neurogenesis. Which neurons get eliminated (random ones

or informational misfits), exactly where in the circuits are they, and what function do they serve? More generally, experiments will be needed to test whether learning makes certain neurons act as repositories for long-term memories—and in so doing, irreversibly inhibits their ability to acquire new information. And it's important to do all these experiments in animals with approximately natural lifestyles: it has been speculated that the reason early primate studies did not spot neurogenesis (Rakic P [1985], Limits of neurogenesis in primates, *Science* 227[4690]) is that laboratory monkeys lead stimulus-impoverished, caged lives. We now know that both stimulating environments and exercise are critical to neurogenesis—exactly what we would expect in the theory of more memories flowing into the system, and therefore more very long-term storage required.

20. Levenson JM, Sweatt JD (2005), Epigenetic mechanisms in memory formation, *Nat Rev Neurosci* 6(2): 108–18. In another example, epigenetic tagging of the genome occurs during consolidation of long-term memories of contextual fear conditioning. In contextual fear conditioning, a noxious stimulus and novel space are paired. The pairing leads to an alteration in the proteins around which DNA wraps and unwraps. The altered genetic expression could accomplish essentially anything, including enhancing synaptic function, the excitability of the neuron, and patterns of receptor expression. When compared with contextual fear conditioning, another form of long-term memory called latent inhibition leads to alterations of a *different* histone, suggesting the possibility of an undiscovered histone code, in which specific types of memory are associated with specific patterns of histone modification.

21. Weaver ICG et al. (2004), Epigenetic programming by maternal behavior, *Nat Neurosci* 7(8): 847. The field of epigenetics examines changes in DNA and the proteins around it that produce lifelong changes in patterns of gene expression. The changes result from an interaction between the genome and the environment. These heritable changes in gene expression are not coded in the DNA sequence itself; this can allow genotypically identical cells to be phenotypically individualized.

22. Brand S (1999), *The Clock of the Long Now: Time and Responsibility* (New York: Basic Books). The idea of pace layering has a history. Brand first created the healthy-civilization diagram with Brian Eno at his studio in London in 1996. Earlier still, in the 1970s, the architect Frank Duffy pointed to four layers in commercial buildings: the set (for example, furniture, which moves often), the scenery (for example, interior walls, which every five to seven years or so are moved around), the services (for instance, the businesses who rent, which modify on a scale of about fifteen years), and the shell (that is, the building itself, which lasts many decades).

23. The counterargument would be that all these other parameters may exist just as a way of keeping homeostasis for the one important change (say, synaptic strengths). To be clear, I think this is unlikely. It would be like pointing to one layer of a society (say, commerce) and arguing that all the other changes in civili-

zation are just a way to keep everything secure so that we always have new places to shop.

24. Neuroscientists typically study this in a way that's not as exciting as falling in love. Instead, they use lab animals, like rats. They teach the rat to perform a task in exchange for reward, and track the speed at which the animal approaches perfect performance. They then extinguish the behavior by taking away the feedback and track how long it takes for the behavior to go away. If they now retrain the animal with feedback, even a very long time later, they find a surprising amount of savings: the animal learns much faster the second time. See Della-Maggiore V, McIntosh AR (2005), Time course of changes in brain activity and functional connectivity associated with long-term adaptation to a rotational transformation, *J Neurophysiol* 93:2254–62; Shadmehr R, Brashers-Krug T (1997), Functional stages in the formation of human long-term motor memory, *J Neurosci* 17:409–19; Landi SM, Baguear F, Della-Maggiore V (2011), One week of motor adaptation induces structural changes in primary motor cortex that predict long-term memory one year later, *J Neurosci* 31:11808–13; Yamamoto K, Hoffman DS, Strick PL (2006), Rapid and longlasting plasticity of input-output mapping, *J Neurophysiol* 96:2797–801.

25. Mulavara AP et al. (2010), Locomotor function after long-duration space flight: Effects and motor learning during recovery, *Exp Brain Res* 202:649–59.

26. Eagleman DM (2011), *Incognito: The Secret Lives of the Brain* (New York: Pantheon). See also Barkow J, Cosmides L, Tooby J (1992), *The Adapted Mind: Evolutionary Psychology and the Generation of Culture* (New York: Oxford University Press).

27. I suggest that the building of the new upon the old underlies the fallibility of eyewitness testimony. Each witness to a crime brings to the table her own history of experience and her own way of understanding the world. Her filters and biases are the sedimentary landscape upon which the new experience lands. It is no surprise that the new input flows down different slopes inside different heads. More generally, the dependence of the present on the past underpins many of the divergences between us, from individual to cultural.

28. Cytowic RE, Eagleman DM (2009), *Wednesday Is Indigo Blue: Discovering the Brain of Synesthesia* (Cambridge, Mass.: MIT Press).

29. Eagleman DM et al. (2007), A standardized test battery for the study of synesthesia, *J Neurosci Methods* 159(1): 139–45. The Synesthesia Battery can be found at synesthete.org.

30. Witthoft N, Winawer J, Eagleman DM (2015), Prevalence of learned graphemecolor pairings in a large online sample of synesthetes, *PLoS One* 10(3): e0118996.

31. We proposed that grapheme-color synesthesia is mental imagery that has been conditioned by experience; that is, it is navigated by your memory. Note this does not contradict findings that the development of the synesthetic response is dependent on genetic predisposition. As for the origin of the colors for the rest of the synesthetes, remember that the magnets were not the only source of external

influence; others ranged from colored alphabets written in books, to alphabet murals, to posters in the homeroom.

32. Plummer W (1997), Total erasure, *People*.

33. Sherry DF, Schacter DL (1987), The evolution of multiple memory systems, *Psychol Rev* 94(4): 439; McClelland JL et al. (1995), Why there are complementary learning systems in the hippocampus and neocortex: Insights from the successes and failures of connectionist models of learning and memory, *Psychol Rev* 102(3): 419.

34. A fast learning rate is necessary for rapid learning; on the other hand, it leads to more interference and to catastrophic failure if trying to store multiple memories. On the flip side, if the rate of changing the strength of connectivity is slow, then the connections come to average over many experiences, and thus simply replicate the underlying statistics of the environment. It used to be thought that the hippocampus "passes" memories through itself and onto the substrate of the cortex, but some newer data suggest that these happen in parallel—both learn at the same time, in parallel. Since the early proposal of this model of complementary learning systems (McCloskey and Cohen [1989]; McClelland et al. [1995]; White [1989]), it has undergone several iterations, all intended to identify where the complementary systems are located in the brain. The original model suggested the hippocampus and cortex (McClelland et al. [1995]), Why there are complementary learning systems in the hippocampus and neocortex, *Psychol Rev* 102:419–57; O'Reilly et al. [2014], Complementary learning systems, *Cogn Sci* 38:1229–48). More recent models have suggested that the different learning rates might happen entirely in the hippocampus: the trisynaptic pathway in CA3 is good at learning clearly demarcated episodes (has a fast learning rate), while the monosynaptic pathway in CA1 is good for statistical learning because of its slower learning rate. See Schapiro et al. (2017), Complementary learning systems within the hippocampus: A neural network modeling approach to reconciling episodic memory with statistical learning, *Phil Trans R Soc B* 372(1711).

11 The Wolf and the Mars Rover

1. Coren MJ (2013), A blind fish inspires new eyes and ears for subs, *FastCoExist*.

2. See, for instance, Leverington M, Shemdin KN (2017), *Principles of Timing in FPGAs*.

3. Eagleman DM (2008), Human time perception and its illusions, *Curr Opin Neurobiol* 18(2): 131–36; Stetson C et al. (2006), Motor-sensory recalibration leads to an illusory reversal of action and sensation, *Neuron* 51(5): 651–59; Parsons B, Novich SD, Eagleman DM (2013), Motor-sensory recalibration modulates perceived simultaneity of cross-modal events, *Front Psychol* 4:46; Cai M, Stetson C, Eagleman DM (2012), A neural model for temporal order judgments and their active recalibration: A common mechanism for space and time?, *Front Psychol* 3:470.

Note that a similar principle is at work when people take out their contact lenses at night and put on a pair of glasses. For the first few moments, their sense of balance is off. Why? Because the glasses warp the scene a bit, such that a movement of the eyes translates to a bigger change in the visual field: the output translates to a slightly unexpected input. How do you solve this rapidly? By swinging your head around for a moment after you put on the glasses. This allows your neural networks to quickly recalibrate the motor output to the sensory input.

4. The examples of smart grids and electrical grids are addressed in more depth in Eagleman DM (2010), *The Safety Net: Avoiding Pandemics and Other Disasters* (Edinburgh: Canongate Books).

12   Finding Ötzi's Long-Lost Love

1. Fowler B (2000), *Iceman: Uncovering the Life and Times of a Prehistoric Man Found in an Alpine Glacier,* (Chicago: U Chicago Press). For a description of the radiology that has been performed, see Gostner P et al. (2011), New radiological insights into the life and death of the Tyrolean Iceman, *Archaeol Sci* 38(12): 3425–31. See also Wierer U et al. (2018), The Iceman's lithic toolkit: Raw material, technology, typology, and use, *PLoS One*; Maixner F et al. (2016), The 5300-year-old Helicobacter pylori genome of the Iceman, *Science* 351(6269): 162–65.

2. Stretesky PB, Lynch MJ (2004), The relationship between lead and crime, *J Health Soc Behav* 45(2): 214–29; Nevin R (2007), Understanding international crime trends: The legacy of preschool lead exposure, *Environ Res* 104(3): 315–36; Reyes JW (2007), Environmental policy as social policy? The impact of childhood lead exposure on crime, *Contrib Econ Anal Pol* 7(1).

FURTHER READING

Ahuja AK et al. (2011). Blind subjects implanted with the Argus II retinal prosthesis are able to improve performance in a spatial-motor task. *Br J Ophthalmol* 95(4): 539–43.

Amedi A, Camprodon J, Merabet L, Meijer P, Pascual-Leone A (2006). Towards closing the gap between visual neuroprostheses and sighted restoration: Insights from studying vision, cross-modal plasticity, and sensory substitution. *J Vision* 6(13): 12.

Amedi A, Floel A, Knecht S, Zohary E, Cohen LG (2004). Transcranial magnetic stimulation of the occipital pole interferes with verbal processing in blind subjects. *Nat Neurosci* 7:1266–70.

Amedi A, Raz N, Azulay H, Malach R, Zohary E (2010). Cortical activity during tactile exploration of objects in blind and sighted humans. *Restor Neurol Neurosci* 28(2): 143–56.

Amedi A, Raz N, Pianka P, Malach R, Zohary E (2003). Early "visual" cortex activation correlates with superior verbal-memory performance in the blind. *Nat Neurosci* 6:758–66.

Amedi A et al. (2007). Shape conveyed by visual-to-auditory sensory substitution activates the lateral occipital complex. *Nat Neurosci* 10:687–89.

Ardouin J et al. (2012). FlyVIZ: A novel display device to provide humans with 360° vision by coupling catadioptric camera with HMD. In *Proceedings of the 18th ACM Symposium on Virtual Reality Software and Technology*.

Arno P, Capelle C, Wanet-Defalque MC, Catalan-Ahumada M, Veraart C (1999). Auditory coding of visual patterns for the blind. *Perception* 28(8): 1013–29.

Arno P et al. (2001). Occipital activation by pattern recognition in the early blind using auditory substitution for vision. *Neuroimage* 13(4): 632–45.

Auvray M, Hanneton S, O'Regan JK (2007). Learning to perceive with a visuo-auditory substitution system: Localisation and object recognition with "The vOICe." *Perception* 36:416–30.

Auvray M, Myin E (2009). Perception with compensatory devices: From sensory substitution to sensorimotor extension. *Cogn Sci* 33(6): 1036–58.

Bach-y-Rita P (2004). Tactile sensory substitution studies. *Ann NY Acad Sci* 1013:83–91.

Bach-y-Rita P, Collins CC, Saunders F, White B, Scadden L (1969). Vision substitution by tactile image projection. *Nature* 221:963–64.

Bach-y-Rita P, Danilov Y, Tyler ME, Grimm RJ (2005). Late human brain plasticity: Vestibular substitution with a tongue BrainPort human-machine interface. *Intellectica* 1(40): 115–22.

Bailey CH, Kandel RR (1993). Structural changes accompanying memory storage. *Ann Rev Physiol* 55:397–426.

Bakin JS, Weinberger NM (1996). Induction of a physiological memory in the cerebral cortex by stimulation of the nucleus basalis. *Proc Natl Acad Sci USA* 93:11219–24.

Bangert M, Schlaug G (2006). Specialization of the specialized in features of external human brain morphology. *Eur J Neurosci* 24:1832–34.

Barinaga M (1992). The brain remaps its own contours. *Science* 258:216–18.

Bear MF, Singer W (1986). Modulation of visual cortical plasticity by acetylcholine and noradrenaline. *Nature* 320:172–76.

Bennett EL, Diamond MC, Krech D, Rosenzweig MR (1964). Chemical and anatomical plasticity of brain. *Science* 164:610–19.

Bliss TV, Lømo T (1973). Long-lasting potentiation of synaptic transmission in the dentate area of the anesthetized rabbit following stimulation of the perforant path. *J Physiol (London)* 232:331–56.

Boldrini M et al. (2018). Human hippocampal neurogenesis persists throughout aging. *Cell Stem Cell* 22(4): 589–99.

Borgstein J, Grootendorst C (2002). Half a brain. *Lancet* 359(9305): 473.

Borsook D et al. (1998). Acute plasticity in the human somatosensory cortex following amputation. *Neuroreport* 9:1013–17.

Bouton CE et al. (2016). Restoring cortical control of functional movement in a human with quadriplegia. *Nature* 533(7602): 247.

Bower TGR (1978). Perceptual development: Object and space. In *Handbook of Perception*, vol. 8, *Perceptual Coding*, edited by EC Carterette and MP Friedman. Academic Press.

Brandt AK and Eagleman DM (2017). *The Runaway Species*. Catapult Press.

Bubic A, Striem-Amit E, Amedi A (2010). Large-scale brain plasticity following blindness and the use of sensory substitution devices. In *Multisensory Object Perception in the Primate Brain*, edited by MJ Naumer and J Kaiser, 351–80. Springer.

Buonomano DV, Merzenich MM (1998). Cortical plasticity: From synapses to maps. *Annu Rev Neurosci* 21:149–86.

Burrone J, O'Byrne M, Murthy VN (2002). Multiple forms of synaptic plasticity triggered by selective suppression of activity in individual neurons. *Nature* 420(6914): 414–18.

Burton H (2003). Visual cortex activity in early and late blind people. *J Neurosci* 23(10): 4005–11.

Burton H, Snyder AZ, Conturo TE, Akbudak E, Ollinger JM, Raichle ME (2002). Adaptive changes in early and late blind: A fMRI study of Braille reading. *J Neurophysiol* 87:589–607.

Cai M, Stetson C, Eagleman DM (2012). A neural model for temporal order judgments and their active recalibration: A common mechanism for space and time? *Front Psychol* 3:470.

Cañón Bermúdez GS, Fuchs H, Bischoff L, Fassbender J, Makarov D (2018). Electronic-skin compasses for geomagnetic field-driven artificial magnetoreception and inter-active electronics. *Nat Electron* 1(11): 589–95.

Carpenter GA, Grossberg S (1987). Discovering order in chaos: Stable self-organization of neural recognition codes. *Ann NY Acad Sci* 504:33–51.

Chebat DR, Harrar V, Kupers R, Maidenbaum S, Amedi A, Ptito M (2018). Sensory sub-stitution and the neural correlates of navigation in blindness. In *Mobility of Visually Impaired People*, 167–200. Springer.

Chorost M (2011). *World Wide Mind: The Coming Integration of Humanity, Machines, and the Internet.* Free Press.

Clark SA, Allard T, Jenkins WM, Merzenich MM (1988). Receptive-fields in the body-surface map in adult cortex defined by temporally correlated inputs. *Nature* 332:444–45.

Cline H (2003). Sperry and Hebb: Oil and vinegar? *Trends Neurosci* 26(12): 655–61.

Cohen LG et al. (1997). Functional relevance of cross-modal plasticity in blind humans. *Nature* 389:180–83.

Collignon O, Lassonde M, Lepore F, Bastien D, Veraart C (2007). Functional cerebral reorganization for auditory spatial processing and auditory substitution of vision in early blind subjects. *Cereb Cortex* 17(2): 457–65.

Collignon O, Renier L, Bruyer R, Tranduy D, Veraart C (2006). Improved selective and divided spatial attention in early blind subjects. *Brain Res* 1075(1): 175–82.

Collignon O, Voss P, Lassonde M, Lepore F (2009). Cross-modal plasticity for the spa-tial processing of sounds in visually deprived subjects. *Exp Brain Res* 192(3): 343–58.

Conner JM, Culberson A, Packowski C, Chiba AA, Tuszynski MH (2003). Lesions of the basal forebrain cholinergic system impair task acquisition and abolish cortical plasticity associated with motor skill learning. *Neuron* 38:819–29.

Constantine-Paton M, Law MI (1978). Eye-specific termination bands in tecta of three-eyed frogs. *Science* 202:639–41.

Cronholm B (1951). Phantom limbs in amputees: A study of changes in the integration of centripetal impulses with special reference to referred sensations. *Acta Psychiatr Neurol Scand Suppl* 72:1–310.

Cronly-Dillon J, Persaud KC, Blore R (2000). Blind subjects construct conscious mental images of visual scenes encoded in musical form. *Proc Biol Sci* 267(1458): 2231–38.

Cronly-Dillon J, Persaud KC, Gregory RP (1999). The perception of visual images encoded in musical form: A study in cross-modality information transfer. *Proc Biol Sci* 266(1436): 2427–33.

Crowley JC, Katz LC (1999). Development of ocular dominance columns in the absence of retinal input. *Nat Neurosci* 2:1125–30.

Cytowic RE, Eagleman DM (2009). *Wednesday Is Indigo Blue: Discovering the Brain of Synesthesia.* MIT Press.

Damasio AR, Tranel D (1993). Nouns and verbs are retrieved with differently distributed neural systems. *Proc Natl Acad Sci USA* 90(11): 4957–60.

D'Angiulli A, Waraich P (2002). Enhanced tactile encoding and memory recognition in congenital blindness. *Int J Rehabil Res* 25(2): 143–45.

Darian-Smith C, Gilbert CD (1994). Axonal sprouting accompanies functional reorganization in adult cat striate cortex. *Nature* 368:737–40.

Day JJ, Sweatt JD (2010). DNA methylation and memory formation. *Nat Neurosci* 13(11): 1319.

Diamond M (2001). Response of the brain to enrichment. *An Acad Bras Ciênc* 73:211–20.

Donati AR et al. (2016). Long-term training with a brain-machine interface-based gait protocol induces partial neurological recovery in paraplegic patients. *Sci Rep* 6:30383.

Dowling J (2008). Current and future prospects for optoelectronic retinal prostheses. *Nature-Eye* 23:1999–2005.

Draganski B, Gaser C, Busch V, Schuierer G, Bogdahn U, May A (2004). Neuroplasticity: Changes in grey matter induced by training. *Nature* 427(6972): 311–12.

Driemeyer J, Boyke J, Gaser C, Büchel C, May A (2008). Changes in gray matter induced by learning—revisited. *PLoS One* 3(7): e2669.

Dudai Y (2004). The neurobiology of consolidations, or, how stable is the engram? *Ann Rev Psychol* 55:51–86.

Eagleman DM (2001). Visual illusions and neurobiology. *Nat Rev Neurosci* 2(12): 920–26.

Eagleman DM (2005). Distortions of time during rapid eye movements. *Nat Neurosci* 8(7): 850–51.

Eagleman DM (2008). Human time perception and its illusions. *Curr Opin Neurobiol* 18(2): 131–36.

Eagleman DM (2009). Silicon immortality: Downloading consciousness into computers. In *This Will Change Everything: Ideas That Will Shape the Future*, edited by J Brockman. Harper Perennial.

Eagleman DM (2010). The strange mapping between the timing of neural signals and perception. In *Issues of Space and Time in Perception and Action*, edited by R Nijhawan. Cambridge University Press.

Eagleman DM (2011). The brain on trial. *Atlantic Monthly*. July/August.

Eagleman DM (2011). *Incognito: The Secret Lives of the Brain*. Pantheon.

Eagleman DM (2012). Synaesthesia in its protean guises. *Br J Psychol* 103(1): 16–19.

Eagleman DM (2015). Can we create new senses for humans? TED talks.

Eagleman DM (2015). *The Brain: The Story of You*. Pantheon.

Eagleman DM, Downar J (2015). *Brain and Behavior: A Cognitive Neuroscience Perspective*. Oxford University Press.

Eagleman DM (2018). We will leverage technology to create new senses. *Wired*.

Eagleman DM, Goodale MA (2009). Why color synesthesia involves more than color. *Trends Cogn Sci* 13(7): 288–92.

Eagleman DM, Jacobson JE, Sejnowski TJ (2004). Perceived luminance depends on temporal context. *Nature* 428(6985): 854.

Eagleman DM, Kagan AD, Nelson SS, Sagaram D, Sarma AK (2007). A standardized test battery for the study of synesthesia. *J Neurosci Methods* 159(1): 139–45.

Eagleman DM, Montague PR (2002). Models of learning and memory. In *Encyclopedia of Cognitive Science*. Macmillan.

Eagleman DM, Pariyadath V (2009). Is subjective duration a signature of coding efficiency? *Philos Trans R Soc* 364(1525): 1841–51.

Eagleman DM, Sejnowski TJ (2000). Motion integration and postdiction in visual awareness. *Science* 287(5460): 2036–38.

Eagleman DM, Vaughn DA (2020). A new theory of dream sleep. Under review.

Edelman GM (1987). *Neural Darwinism: The Theory of Neuronal Group Selection*. Basic Books.

Elbert T, Pentev C, Wienbruch C, Rockstroh B, Taub E (1995). Increased finger representation of the fingers of the left hand in string players. *Science* 270:305–6.

Elbert T, Rockstroh B (2004). Reorganization of human cerebral cortex: The range of changes following use and injury. *Neuroscientist* 10:129–41.

Eriksson PS et al. (1998). Neurogenesis in the adult human hippocampus. *Nat Med* 4(11): 1313–17.

Feuillet L, Dufour H, Pelletier J (2007). Brain of a white-collar worker. *Lancet* 370:262.

Finney EM, Fine I, Dobkins KR (2001). Visual stimuli activate auditory cortex in the deaf. *Nat Neurosci* 4(12): 1171–73.

Flor H, Elbert T, Knecht S, Wienbruch C, Pantev C, Birbaumer N, Larbig W, Taub E (1995). Phantom-limb pain as a perceptual correlate of cortical reorganization following arm amputation. *Nature* 375(6531): 482–84.

Florence SL, Taub HB, Kaas JH (1998). Large-scale sprouting of cortical connections after peripheral injury in adult macaque monkeys. *Science* 282:1117–21.

Fuhr P, Cohen LG, Dang N, Findley TW, Haghighi S, Oro J, Hallett M (1992). Physiological analysis of motor reorganization following lower limb amputation. *Electroencephalogr Clin Neurophysiol* 85(1): 53–60.

Fusi S, Drew PJ, Abbott LF (2005). Cascade models of synaptically stored memories. *Neuron* 45(4): 599–611.

Gougoux F, Lepore F, Lassonde M, Voss P, Zatorre RJ, Belin P (2004). Neuropsychology: Pitch discrimination in the early blind. *Nature* 430(6997): 309.

Gougoux F, Zatorre RJ, Lassonde M, Voss P, Lepore F (2005). A functional neuroimaging study of sound localization: Visual cortex activity predicts performance in early-blind individuals. *PLoS Biol* 3(2): e27.

Gould E, Beylin AV, Tanapat P, Reeves A, Shors TJ (1999). Learning enhances adult neurogenesis in the hippocampal formation. *Nat Neurosci* 2:260–65.

Gould E, Reeves A, Graziano MSA, Gross C (1999). Neurogenesis in the neocortex of adult primates. *Science* 286:548–52.

Gu Q (2003). Contribution of acetylcholine to visual cortex plasticity. *Neurobiol Learn Mem* 80:291–301.

Hallett M (1999). Plasticity in the human motor system. *Neuroscientist* 5:324–32.

Halligan PW, Marshall JC, Wade DT (1994). Sensory disorganization and perceptual plasticity after limb amputation: A follow-up study. *Neuroreport* 5:1341–45.

Hamilton RH, Keenan JP, Catala MD, Pascual-Leone A (2000). Alexia for Braille following bilateral occipital stroke in an early blind woman. *Neuroreport* 11:237–40.

Hamilton RH, Pascual-Leone A, Schlaug G (2004). Absolute pitch in blind musicians. *Neuroreport* 15:803–6.

Hasselmo ME (1995). Neuromodulation and cortical function: Modeling the physiological basis of behavior. *Behav Brain Res* 67:1–27.

Hawkins J, Blakeslee S (2004). *On Intelligence*. Times Books.

Hochberg LR, Serruya MD, Friehs GM, Mukand JA, Saleh M, Caplan AH, Branner A, Chen D, Penn RD, Donoghue JP (2006). Neuronal ensemble control of prosthetic devices by a human with tetraplegia. *Nature* 442:164–71.

Hoffman KL, McNaughton BL (2002). Coordinated reactivation of distributed memory traces in primate neocortex. *Science* 297:2070.

Hoffmann R et al. (2018). Evaluation of an audio-haptic sensory substitution device for enhancing spatial awareness for the visually impaired. *Optom Vis Sci* 95(9): 757.

Hubel DH, Wiesel TN (1965). Binocular interaction in striate cortex of kittens reared with artificial squint. *J Neurophysiol* 28:1041–59.

Hurovitz C, Dunn S, Domhoff GW, Fiss H (1999). The dreams of blind men and women: A replication and extension of previous findings. *Dreaming* 9:183–93.

Jacobs GH, Williams GA, Cahill H, Nathans J (2007). Emergence of novel color vision in mice engineered to express a human cone photopigment. *Science* 315(5819): 1723–25.

Jameson KA (2009). Tetrachromatic color vision. In *The Oxford Companion to Consciousness*, edited by P Wilken, T Bayne, and A Cleeremans. Oxford University Press.

Johnson JS, Newport EL (1989). Critical period effects in second language learning: The influence of maturational state on the acquisition of English as a second language. *Cogn Psychol* 21:60–99.

Jones EG (2000). Cortical and subcortical contributions to activity-dependent plasticity in primate somatosensory cortex. *Annu Rev Neurosci* 23:1–37.

Jones EG, Pons TP (1998). Thalamic and brainstem contributions to large-scale plasticity of primate somatosensory cortex. *Science* 282(5391): 1121–25.

Karl A, Birbaumer N, Lutzenberger W, Cohen LG, Flor H (2001). Reorganization of motor and somatosensory cortex in upper extremity amputees with phantom limb pain. *J Neurosci* 21:3609–18.

Karni A, Meyer G, Jezzard P, Adams M, Turner R, Ungerleider L (1995). Functional MRI evidence for adult motor cortex plasticity during motor skill learning. *Nature* 377:155–58.

Kay L (2000). Auditory perception of objects by blind persons, using a bioacoustic high resolution air sonar. *J Acoust Soc Am* 107(6): 3266–76.

Kennedy PR, Bakay RA (1998). Restoration of neural output from a paralyzed patient by a direct brain connection. *Neuroreport* 9:1707–11.

Kilgard MP, Merzenich MM (1998). Cortical map reorganization enabled by nucleus basalis activity. *Science* 279:1714–18.

Knudsen EI (2002). Instructed learning in the auditory localization pathway of the barn owl. *Nature* 417:322–28.

Kubanek M, Bobulski J (2018). Device for acoustic support of orientation in the surroundings for blind people. *Sensors* 18(12): 4309.

Kuhl PK (2004). Early language acquisition: Cracking the speech code. *Nat Rev Neurosci* 5:831–43.

Kupers R, Ptito M (2014). Compensatory plasticity and cross-modal reorganization following early visual deprivation. *Neurosci Biobehav Rev* 41:36–52.

Law MI, Constantine-Paton M (1981). Anatomy and physiology of experimentally produced striped tecta. *J Neurosci* 1:741–59.

Lenay C, Gapenne O, Hanneton S, Marque C, Genouëlle C (2003). Sensory substitution: Limits and perspectives. In *Touching for Knowing, Cognitive Psychology of Haptic Manual Perception*, edited by Y Hatwell, A Streri, and E Gentaz, 275–92. John Benjamins.

Levy B (2008). The blind climber who "sees" with his tongue. *Discover*, June 22, 2008.

Lisman J, Cooper K, Sehgal M, Silva AJ (2018). Memory formation depends on both synapse-specific modifications of synaptic strength and cell-specific increases in excitability. *Nat Neurosci* 12:1.

Lobo L et al. (2018). Sensory substitution: Using a vibrotactile device to orient and walk to targets. *J Exp Psychol Appl* 24(1): 108.

Macpherson F, ed. (2018). *Sensory Substitution and Augmentation*. Oxford University Press.

Mancuso K, Hauswirth WW, Li Q, Connor TB, Kuchenbecker JA, Mauck MC, Neitz J, Neitz M (2009). Gene therapy for red-green colour blindness in adult primates. *Nature* 461:784–88.

Maravita A, Iriki A (2004). Tools for the body (schema). *Trends Cogn Sci* 8:79–86.

Martin SJ, Grimwood PD, Morris RG (2000). Synaptic plasticity and memory: An evaluation of the hypothesis. *Annu Rev Neurosci* 23:649–711.

Massiceti D, Hicks SL, van Rheede JJ (2018). Stereosonic vision: Exploring visual-to-auditory sensory substitution mappings in an immersive virtual reality navigation paradigm. *PLoS One* 13(7).

Matteau I, Kupers R, Ricciardi E, Pietrini P, Ptito M (2010). Beyond visual, aural, and haptic movement perception: hMT+ is activated by electrotactile motion stimulation of the tongue in sighted and in congenitally blind individuals. *Brain Res Bull* 82(5–6): 264–70.

Meijer PB (1992). An experimental system for auditory image representations. *IEEE Trans Biomed Eng* 39(2): 112–21.

Merabet LB et al. (2007). Combined activation and deactivation of visual cortex during tactile sensory processing. *J Neurophysiol* 97:1633–41.

Merabet LB et al. (2008). Rapid and reversible recruitment of early visual cortex for touch. *PLoS One* 3(8): e3046.

Merabet LB, Pascual-Leone A (2010). Neural reorganization following sensory loss: The opportunity of change. *Nat Rev Neurosci* 11(1): 44–52.

Merabet LB, Rizzo J, Amedi A, Somers D, Pascual-Leone A (2005). What blindness can tell us about seeing again: Merging neuroplasticity and neuroprostheses. *Nat Rev Neurosci* 6(1): 71–77.

Merzenich MM (1998). Long-term change of mind. *Science* 282(5391): 1062–63.

Merzenich MM et al. (1984). Somatosensory cortical map changes following digit amputation in adult monkeys. *J Comp Neurol* 224:591–605.

Miller TC, Crosby TW (1979). Musical hallucinations in a deaf elderly patient. *Ann Neurol* 5:301–2.

Mitchell SW (1872). *Injuries of Nerves and Their Consequences.* Lippincott.

Montague PR, Eagleman DM, McClure SM, Berns GS (2002). Reinforcement learning. In *Encyclopedia of Cognitive Science.* Macmillan.

Moosa AN et al. (2013). Long-term functional outcomes and their predictors after hemispherectomy in 115 children. *Epilepsia* 54(10): 1771–79.

Muckli L, Naumer MJ, Singer W (2009). Bilateral visual field maps in a patient with only one hemisphere. *Proc Natl Acad Sci USA* 106(31): 13034–39.

Muhlau M et al. (2006). Structural brain changes in tinnitus. *Cereb Cortex* 16:1283–88.

Nagel SK, Carl C, Kringe T, Märtin R, König P (2005). Beyond sensory substitution—learning the sixth sense. *J Neural Eng* 2(4): R13–26.

Nau AC, Pintar C, Arnoldussen A, Fisher C (2015). Acquisition of visual perception in blind adults using the BrainPort artificial vision device. *Am J Occup Ther* 69(1): 1–8.

Neely RM, Piech DK, Santacruz SR, Maharbiz MM, Carmena JM (2018). Recent advances in neural dust: Towards a neural interface platform. *Curr Opin Neurobiol* 50:64–71.

Neville H, Bavelier D (2002). Human brain plasticity: Evidence from sensory deprivation and altered language experience. *Prog Brain Res* 138:177–88.

Noë A (2009). *Out of Our Heads.* Hill and Wang.

Norimoto H, Ikegaya Y (2015). Visual cortical prosthesis with a geomagnetic compass restores spatial navigation in blind rats. *Curr Biol* 25(8): 1091–95.

Nottebohm F (2002). Neuronal replacement in adult brain. *Brain Res Bull* 57(6): 737–49.

Novich SD, Eagleman DM (2015). Using space and time to encode vibrotactile information: Toward an estimate of the skin's achievable throughput. *Exp Brain Res* 233(10): 2777–88.

Nudo RJ, Milliken GW, Jenkins WM, Merzenich MM (1996). Use-dependent alterations of movement representations in primary motor cortex of adult squirrel monkeys. *J Neurosci* 16(2): 785–807.

O'Brien J, Bloomfield SA (2018). Plasticity of retinal gap junctions: Roles in synaptic physiology and disease. *Annu Rev Vis Sci* 4:79–100.

O'Regan JK, Noë A (2001). A sensorimotor account of vision and visual consciousness. *Behav Brain Sci* 24(5): 939–73; discussion 973–1031.

Orsetti M, Casamenti F, Pepeu G (1996). Enhanced acetylcholine release in the hippocampus and cortex during acquisition of an operant behavior. *Brain Res* 724: 89–96.

Ortiz-Terán L et al. (2016). Brain plasticity in blind subjects centralizes beyond the modal cortices. *Front Syst Neurosci* 10:61.

Ortiz-Terán L et al. (2017). Brain circuit–gene expression relationships and neuroplasticity of multisensory cortices in blind children. *Proc Natl Acad Sci* 114(26): 6830–35.

Osinski D, Hjelme DR (2018). A sensory substitution device inspired by the human visual system. *11th International Conference on Human System Interaction.*

Parsons B, Novich SD, Eagleman DM (2013). Motor-sensory recalibration modulates perceived simultaneity of cross-modal events. *Front Psychol* 4:46.

Pascual-Leone A, Amedi A, Fregni F, Merabet L (2005). The plastic human brain cortex. *Annu Rev Neurosci* 28:377–401.

Pascual-Leone A, Hamilton R (2001). The metamodal organization of the brain. In *Vision: From Neurons to Cognition,* edited by C Casanova and M Ptito, 427–45. Elsevier Science.

Pascual-Leone A, Peris M, Tormos JM, Pascual AP, Catala MD (1996). Reorganization of human cortical motor output maps following traumatic forearm amputation. *Neuroreport* 7:2068–70.

Pasupathy A, Miller EK (2005). Different time courses of learning-related activity in the prefrontal cortex and striatum. *Nature* 433(7028): 873–76.

Penfield W (1961). Activation of the record of human experience. *Ann R Coll Surg Engl* 29(2): 77–84.

Perrett SP, Dudek SM, Eagleman DM, Montague PR, Friedlander MJ (2001). LTD induction in adult visual cortex: Role of stimulus timing and inhibition. *J Neurosci* 21(7): 2308–19.

Petitto LA, Marentette PF (1991). Babbling in the manual mode: Evidence for the ontogeny of language. *Science* 251:1493–96.

Poirier C, De Volder AG, Scheiber C (2007). What neuroimaging tells us about sensory substitution. *Neurosci Biobehav Rev* 31:1064–70.

Pons TP, Garraghty PE, Ommaya AK, Kaas JH, Taub E, Mishkin M (1991). Massive cortical reorganization after sensory deafferentation in adult macaques. *Science* 252:1857–60.

Proulx MJ, Stoerig P, Ludowig E, Knoll I (2008). Seeing "where" through the ears: Effects of learning-by-doing and long-term sensory deprivation on localization based on image-to-sound substitution. *PLoS One* 3(3): e1840.

Ptito M, Fumal A, De Noordhout AM, Schoenen J, Gjedde A, Kupers R (2008). TMS of the occipital cortex induces tactile sensations in the fingers of blind Braille readers. *Exp Brain Res* 184(2): 193–200.

Rajangam S, Tseng PH, Yin A, Lehew G, Schwarz D, Lebedev MA, Nicolelis MA (2016). Wireless cortical brain-machine interface for whole-body navigation in primates. *Sci Rep* 6:22170.

Ramachandran VS (1993). Behavioral and MEG correlates of neural plasticity in the adult human brain. *Proc Natl Acad Sci USA* 90:10413–20.

Ramachandran VS, Rogers-Ramachandran D, Stewart M (1992). Perceptual correlates of massive cortical reorganization. *Science* 258:1159–60.

Rao RP, Sejnowski TJ (2003). Self-organizing neural systems based on predictive learning. *Philos Transact A Math Phys Eng Sci* 361(1807): 1149–75.

Raz N, Amedi A, Zohary E (2005). V1 activation in congenitally blind humans is associated with episodic retrieval. *Cereb Cortex* 15:1459–68.

Renier L, Anurova I, De Volder AG, Carlson S, VanMeter J, Rauschecker JP (2010). Preserved functional specialization for spatial processing in the middle occipital gyrus of the early blind. *Neuron* 68(1): 138–48.

Renier L, De Volder AG, Rauschecker JP (2014). Cortical plasticity and preserved function in early blindness. *Neurosci Biobehav Rev* 41:53–63.

Ribot T (1882). *Diseases of the Memory: An Essay in the Positive Psychology.* D. Appleton.

Roberson ED, Sweatt JD (1999). A biochemical blueprint for long-term memory. *Learn Mem* 6(4): 381–88.

Rosenzweig MR, Bennett EL (1996). Psychobiology of plasticity: Effects of training and experience on brain and behavior. *Behav Brain Res* 78:57–65.

Royer S, Pare D (2003). Conservation of total synaptic weight through balanced synaptic depression and potentiation. *Nature* 422(6931): 518–22.

Sachdev RNS, Lu SM, Wiley RG, Ebner FF (1998). Role of the basal forebrain cholinergic projection in somatosensory cortical plasticity. *J Neurophysiol* 79:3216–28.

Sadato N, Pascual-Leone A, Grafman J, Deiber MP, Ibanez V, Hallett M (1998). Neural networks for Braille reading by the blind. *Brain* 121:1213–29.

Sampaio E, Maris S, Bach-y-Rita P (2001). Brain plasticity: "Visual" acuity of blind persons via the tongue. *Brain Res* 908(2): 204–7.

Sathian K, Stilla R (2010). Cross-modal plasticity of tactile perception in blindness. *Restor Neurol Neurosci* 28(2): 271–81.

Schulz LE, Gopnik A (2004). Causal learning across domains. *Dev Psychol* 40:162–76.

Schweighofer N, Arbib MA (1998). A model of cerebellar metaplasticity. *Learn Mem* 4(5): 421–28.

Sharma J, Angelucci A, Sur M (2000). Induction of visual orientation modules in auditory cortex. *Nature* 404:841–47.

Simon M (2019). How I became a robot in London—from 5,000 miles away. *Wired.*

Singh AK, Phillips F, Merabet LB, Sinha P (2018). Why does the cortex reorganize after sensory loss? *Trends Cogn Sci* 22(7): 569–82.

Smirnakis SM, Brewer AA, Schmid MC, Tolias AS, Schüz A, Augath M, Inhoffen W, Wandell BA, Logothetis NK (2005). Lack of long-term cortical reorganization after macaque retinal lesions. *Nature* 435(7040): 300–307.

Southwell DG, Froemke RC, Alvarez-Buylla A, Stryker MP, Gandhi SP (2010). Cortical plasticity induced by inhibitory neuron transplantation. *Science* 327(5969): 1145–48.

Spedding M, Gressens P (2008). Neurotrophins and cytokines in neuronal plasticity. *Novartis Found Symp* 289:222–33; discussion 233–40.

Steele CJ, Zatorre RJ (2018). Practice makes plasticity. *Nat Neurosci* 21(12): 1645.

Stetson C, Cui X, Montague PR, Eagleman DM (2006). Motor-sensory recalibration leads to an illusory reversal of action and sensation. *Neuron* 51(5): 651–59.

Tapu R, Mocanu B, Zaharia T (2018). Wearable assistive devices for visually impaired: A state of the art survey. *Pattern Recognit Lett.*

Thaler L, Goodale MA (2016). Echolocation in humans: An overview. *Wiley Interdiscip Rev Cogn Sci* 7(6): 382–93.

Thiel CM, Friston KJ, Dolan RJ (2002). Cholinergic modulation of experience-dependent plasticity in human auditory cortex. *Neuron* 35:567–74.

Tulving E, Hayman CAG, Macdonald CA (1991). Long-lasting perceptual priming and semantic learning in amnesia: A case experiment. *J Exp Psychol Learn Mem Cogn* 17:595–617.

Udin SH (1977). Rearrangements of the retinotectal projection in *Rana pipiens* after unilateral caudal half-tectum ablation. *J Comp Neurol* 173:561–82.

Velliste M, Perel S, Spalding MC, Whitford AS, Schwartz AB (2008). Cortical control of a prosthetic arm for self-feeding. *Nature* 453:1098–101.

von Melchner L, Pallas SL, Sur M (2000). Visual behaviour mediated by retinal projections directed to the auditory pathway. *Nature* 404:871–76.

Voss P, Gougoux F, Lassonde M, Zatorre RJ, Lepore F (2006). A positron emission tomography study during auditory localization by late-onset blind individuals. *Neuroreport* 17(4): 383–88.

Voss P, Gougoux F, Zatorre RJ, Lassonde M, Lepore F (2008). Differential occipital responses in early- and late-blind individuals during a sound-source discrimination task. *Neuroimage* 40(2): 746–58.

Weaver IC et al. (2004). Epigenetic programming by maternal behavior. *Nat Neurosci* 7(8): 847–54.

Weiss T, Miltner WH, Liepert J, Meissner W, Taub E (2004). Rapid functional plasticity in the primary somatomotor cortex and perceptual changes after nerve block. *Eur J Neurosci* 20:3413–23.

Whitlock JR, Heynen AJ, Shuler MG, Bear MF (2006). Learning induces long-term potentiation in the hippocampus. *Science* 313(5790): 1093–97.

Wiesel TN, Hubel DH (1963). Single-cell responses in striate cortex of kittens deprived of vision in one eye. *J Neurophysiol* 26:1003–17.

Witthoft N, Winawer J, Eagleman DM (2015). Prevalence of learned grapheme-color pairings in a large online sample of synesthetes. *PLoS One* 10(3).

Won AS, Bailenson JN, Lanier J (2015). Homuncular flexibility: The human ability to inhabit nonhuman avatars. In *Emerging Trends in the Social and Behavioral Sciences*, edited by R Scott and M Buchmann, 1–6. John Wiley & Sons.

Yamahachi H, Marik SA, McManus JN, Denk W, Gilbert CD (2009). Rapid axonal sprouting and pruning accompany functional reorganization in primary visual cortex. *Neuron* 64(5): 719–29.

Yang TT, Gallen CC, Ramachandran VS, Cobb S, Schwartz BJ, Bloom FE (1994). Noninvasive detection of cerebral plasticity in adult human somatosensory cortex. *Neuroreport* 5:701–4.

Zola-Morgan SM, Squire LR (1990). The primate hippocampal formation: Evidence for a time-limited role in memory storage. *Science* 250(4978): 288–90.

INDEX

An "n" references an endnote number on that page.

Page

6 Reprinted from Kliemann, D., et al. (2019). Intrinsic functional connectivity of the brain in adults with a single cerebral hemisphere. *Cell Reports.* 2019 Nov 19; 29(8): 2398–407. Copyright © 2019, with permission from Elsevier.

23 Melissa Lyttle/*Tampa Bay Times*

27 Courtesy of the author

32 Courtesy of the author

35 Courtesy of the author

37 Courtesy of the author

46 Courtesy of the author

54 Javier Fadul, Kara Gray, and Culture Pilot

56 Courtesy of the author

57 Javier Fadul, Kara Gray, and Culture Pilot

58 Sharon Steinmann/AL.com/*The Birmingham News*

59 Anthony Souffle/*Chicago Tribune*/Getty Images *(top)*

59 Courtesy of KTTC News *(bottom)*

62 Javier Fadul, Kara Gray, and Culture Pilot

63 Javier Fadul, Kara Gray, and Culture Pilot

68 Courtesy of the author

69 Javier Fadul, Kara Gray, and Culture Pilot

70 Javier Fadul, Kara Gray, and Culture Pilot

72 Ted West/Hulton Archive/Getty Images

77 Syed Rahman

79 Syed Rahman and Emily Stevens

82 Courtesy of the author

85 Lars Norgaard

88 Jérôme Ardouin

98 Courtesy of the author

108 Javier Fadul, Kara Gray, and Culture Pilot

111 Mukhopadhyay B, Shukla RM, Mukhopadhyay M, Mandal KC, Haldar P, Benare A (2012). Spectrum of human tails: A report of six cases. *Journal*

of Indian Association of Pediatric Surgeons, 17(1), 23–25. https://doi.org/10.4103/0971-9261.91082.
112 Associated Press
113 USA Archery
114 Atort Photography
115 Fabian Lewkowicz *(top);* Lionel Hahn/Sipa USA *(bottom)*
117 Destin Sandlin
119 Viktor Zykov/Creative Machines Lab, Columbia University
125 Andrew B. Schwartz
126 Gregoire Cirade/Science Photo Library
128 Courtesy of the author
129 Andrew B. Schwartz
141 Courtesy of the author
151 D. Eagleman and J. Downar, *Brain and Behavior,* Oxford University Press
161 Courtesy of the author
162 IBM
165 Paul Parker/Science Photo Library
175 Courtesy of the author
176 Courtesy of the author
177 Courtesy of the author
179 Courtesy of the author
183 D. Eagleman and J. Downar, *Brain and Behavior,* Oxford University Press
184 Wikipedia, Creative Commons Attribution-Share Alike 4.0 International license
200 Nina Leen/Getty Images
209 Guy Pearce in Memento © Summit Entertainment
220 Courtesy of the author
226 Witthoft N, Winawer J, Eagleman DM (2015) Prevalence of Learned Grapheme-Color Pairings in a Large Online Sample of Synesthetes. *PLoS ONE* 10(3): e0118996. https://doi.org/10.1371/journal.pone.0118996
228 Courtesy of the author